Nuclear Science, Technology, and Civilization

AF622606

Nuclear Science, Technology, and Civilization

Edited by

Tao Zhou
Department of Nuclear Science and Technology, Institute of Nuclear Thermal-Hydraulic Safety and Standardization, Southeast University, Nanjing, Jiangsu, China

Yao Yao
Department of Nuclear Science and Technology, Institute of Nuclear Thermal-Hydraulic Safety and Standardization, Southeast University, Nanjing, Jiangsu, China

Shang Mao
School of Energy and Power Engineering, Zhengzhou University of Light Industry, Zhengzhou, Henan, China

ELSEVIER

Elsevier
Radarweg 29, PO Box 211, 1000 AE Amsterdam, Netherlands
50 Hampshire Street, 5th Floor, Cambridge, MA 02139, United States

Copyright © 2026 Elsevier Inc. All rights are reserved, including those for text and data mining, AI training, and similar technologies.

Publisher's note: Elsevier takes a neutral position with respect to territorial disputes or jurisdictional claims in its published content, including in maps and institutional affiliations.

Books and Journals published by Elsevier comply with applicable product safety requirements. For any product safety concerns or queries, please contact our authorised representative, Elsevier B.V., at productsafety@elsevier.com.

For accessibility purposes, images in electronic versions of this book are accompanied by alt text descriptions provided by Elsevier. For more information, see https://www.elsevier.com/about/accessibility.

No part of this publication may be reproduced or transmitted in any form or by any means, electronic or mechanical, including photocopying, recording, or any information storage and retrieval system, without permission in writing from the publisher. Details on how to seek permission, further information about the Publisher's permissions policies and our arrangements with organizations such as the Copyright Clearance Center and the Copyright Licensing Agency, can be found at our website: www.elsevier.com/permissions.

This book and the individual contributions contained in it are protected under copyright by the Publisher (other than as may be noted herein).

Notices
Knowledge and best practice in this field are constantly changing. As new research and experience broaden our understanding, changes in research methods, professional practices, or medical treatment may become necessary.

Practitioners and researchers must always rely on their own experience and knowledge in evaluating and using any information, methods, compounds, or experiments described herein. In using such information or methods they should be mindful of their own safety and the safety of others, including parties for whom they have a professional responsibility.

To the fullest extent of the law, neither the Publisher nor the authors, contributors, or editors, assume any liability for any injury and/or damage to persons or property as a matter of products liability, negligence or otherwise, or from any use or operation of any methods, products, instructions, or ideas contained in the material herein.

ISBN: 978-0-443-13941-3

For Information on all Elsevier publications visit our website at https://www.elsevier.com/books-and-journals

Publisher: Priscilla Braglia
Acquisitions Editor: Priscilla Braglia
Editorial Project Manager: Juwairiya Ali
Production Project Manager: Maria Shalini A
Cover Designer: Mark Rogers

Working together to grow libraries in developing countries
www.elsevier.com • www.bookaid.org

Typeset by Aptara, New Delhi, India

Contents

Contributors

Zongyan Bai, Department of Nuclear Science and Technology, Institute of Nuclear Thermal-hydraulic Safety and Standardization, Southeast University, Nanjing, Jiangsu, China

Shilei Dun, Department of Nuclear Science and Technology, Institute of Nuclear Thermal-hydraulic Safety and Standardization, Southeast University, Nanjing, Jiangsu, China

Tianyu Gao, Department of Nuclear Science and Technology, Institute of Nuclear Thermal-hydraulic Safety and Standardization, Southeast University, Nanjing, Jiangsu, China

Dongli Huang, Department of Nuclear Science and Technology, Institute of Nuclear Thermal-hydraulic Safety and Standardization, Southeast University, Nanjing, Jiangsu, China

Chunmei Liu, Department of Nuclear Science and Technology, Institute of Nuclear Thermal-Hydraulic Safety and Standardization, Southeast University, Nanjing, Jiangsu, China

Peng Liu, Department of Nuclear Science and Technology, Institute of Nuclear Thermal-hydraulic Safety and Standardization, Southeast University, Nanjing, Jiangsu, China

Wenbin Liu, Department of Nuclear Science and Technology, Institute of Nuclear Thermal-hydraulic Safety and Standardization, Southeast University, Nanjing, Jiangsu, China

Xiaofang Liu, Department of Nuclear Science and Technology, Institute of Nuclear Thermal-Hydraulic Safety and Standardization, Southeast University, Nanjing, Jiangsu, China

Huaichang Lu, Department of Nuclear Science and Technology, Institute of Nuclear Thermal-Hydraulic Safety and Standardization, Southeast University, Nanjing, Jiangsu, China

Shang Mao, School of Energy and Power Engineering, Zhengzhou University of Light Industry, Zhengzhou, Henan, China

Jianyu Tang, Department of Nuclear Science and Technology, Institute of Nuclear Thermal-hydraulic Safety and Standardization, Southeast University, Nanjing, Jiangsu, China

Dong Wei, Department of Nuclear Science and Technology, Institute of Nuclear Thermal-hydraulic Safety and Standardization, Southeast University, Nanjing, Jiangsu, China

Peng Xu, Institute of Advanced Nuclear Energy Technology, State Power Investment Corporation Research Institute, Beijing, Beijing, China

Chunhui Xue, Department of Nuclear Science and Technology, Institute of Nuclear Thermal-hydraulic Safety and Standardization, Southeast University, Nanjing, Jiangsu, China

Yao Yao, Department of Nuclear Science and Technology, Institute of Nuclear Thermal-Hydraulic Safety and Standardization, Southeast University, Nanjing, Jiangsu, China

Haolei Zhang, Department of Nuclear Science and Technology, Institute of Nuclear Thermal-hydraulic Safety and Standardization, Southeast University, Nanjing, Jiangsu, China

Tao Zhou, Department of Nuclear Science and Technology, Institute of Nuclear Thermal-Hydraulic Safety and Standardization, Southeast University, Nanjing, Jiangsu, China

List of authors

Author's brief introduction

Pro. Tao Zhou: Professor in School of Energy and Environment of Southeast University, the director of Department of Nuclear Science and Technology, a doctoral supervisor. He has many years of experience in various national fund vertical projects and major horizontal projects, and has accumulated rich research experience. He has published more than 100 papers at home and abroad, including more than 100 SCI and EI indexed papers. He has more than 100 invention patents and copyrights, edited 4 books and participated in 2 books.

Yao Yao: Ph.D. candidate of Department of Nuclear Science and Technology.

Shang Mao: Ph.D. of Department of Nuclear Science and Technology. The research direction is particulate deposition. He pays close attention to the energy conservation, emission reduction and environmental protection.

Peng Xu: Ph.D. of Department of Nuclear Science and Technology. The research direction is ocean nuclide migration, and he can think about the impact of nuclear energy on human beings from multiple perspectives.

Chunmei Liu: Ph.D. of Economics, the research direction is economic analysis, especially economic analysis of urban construction in combination with land, water, climate and policies. The multi-angle dialectical analysis of cases can be performed.

Dongli Huang: Associate researcher, Ph.D. of Nuclear Engineering at Purdue University.

Da Chen: Associate researcher, Ph.D. at City University of Hong Kong.

Biao Zhang: Associate Professor in School of Energy and Environment of Southeast University.

Chunhui Xue: Ph.D. of Nuclear Engineering.

Wenbin Liu: Studies on the impact of COVID-19 on the operation of nuclear power plants.

Dong Wei, Yitong Zhang, Jianyu Tang, Huaichang Lu, Tianyu Gao, Xiaofang Liu, Peng Liu, Haolei Zhang, Shilei Dun, Zongyan Bai, Shuyu He, Bokang Chang: Nuclear science and technology postgraduates. They learned the knowledge of Nuclear physics and nuclear radiation, Nuclear turbine and thermal system, Reactor control and nuclear power plant instrumentation system, nuclear reactor measuring principle, nuclear reactor physics, nuclear reactor thermo-hydraulics, nuclear power plant system and equipment, nuclear power plant nuclear safety, the operation of the pressurized water reactor nuclear power plant. They are familiar with the related concepts of nuclear science and technology.

	Preface	Chapter 1	Chapter 2	Chapter 3	Chapter 4	Chapter 5	Chapter 6	Chapter 7	Chapter 8
Tao Zhou	√	√*	√	√*	√	√	√*	√	√
Yao Yao	√*	√	√*	√	√*			√	
Shang Mao				√		√*	√	√*	√*
Dongli Huang					√				
Chunhui Xue					√			√	
Chunmei Liu				√		√			
Peng Xu		√							√
Wenbin Liu			√	√		√		√	
Dong Wei							√		√
Jianyu Tang					√	√		√	
Huaichang Lu		√					√		
Tianyu Gao			√				√		
Xiaofang Liu		√		√					
Peng Liu					√				
Haolei Zhang			√	√					√
Shilei Dun			√	√	√				
Zongyan Bai					√				

Note: * is the person chiefly in charge of this chapter

Preface

Guided by the concept of nuclear civilization, this book aims to summarize and present the impact of nuclear science and technology on human civilization and social development from diverse perspectives. It seeks to help readers understand the history, related technologies, and applications of nuclear science and technology, as well as their influence across different regions, populations, professional fields, and contexts. In a manner consistent with science popularization, it examines the relationship between nuclear science and technology and human civilization. In the face of increasingly severe global climate change, reducing high-carbon emissions and optimizing industrial and energy structures have become urgent imperatives. Countries around the world are actively taking measures to mitigate global warming. As a clean, low-carbon, efficient, economical, and stable energy source, nuclear energy is being considered for development and deployment by many nations. For example, the United States has renewed its emphasis on nuclear energy development; the European Union has reclassified nuclear energy and natural gas as green energy sources; and China, under its "dual carbon" goals, advocates for the active and orderly development of nuclear power. As environmental protection requirements intensify, nuclear science and technology are poised to assume an even more significant role. In addition to actively responding to environmental imperatives such as the "dual carbon" goals, nuclear science and technology will achieve mutually empowering development driven by new quality productive forces—including artificial intelligence—and will also play a uniquely important role under exceptional circumstances, such as epidemic prevention and control exemplified by the global COVID-19 pandemic. Nevertheless, severe accidents at nuclear power plants may release radioactivity, leading to public concern and fear—a case in point being the planned discharge of Fukushima's nuclear-contaminated water into the sea following the Fukushima nuclear accident. Although researchers in nuclear-related fields generally exhibit less anxiety and fear regarding nuclear issues, such apprehension may persist among the general public, policymakers, and even nonnuclear researchers. Consequently, this book adopts a dialectical perspective to objectively synthesize and analyze the influence of nuclear science and technology development on human civilization. It covers representative developments in nuclear science and technology in leading nuclear power countries—such as the United States, China, Europe, and Russia—while also illustrating the relationship between the energy economy and public security, safety, and prosperity. Upon reading this book, readers will gain a deeper understanding of nuclear science and technology, assess its development—including nuclear energy—from an objective and dialectical standpoint, and accurately comprehend its safety and its developmental value to human civilization.

This book can serve as either a reference or training material for students specializing in nuclear science and engineering. It consists of eight chapters: Chapter 1 discusses the definition, role, and classification of nuclear science and technology; Chapter 2 examines the history and evolution of nuclear science and technology; Chapter 3 explores the impact of nuclear science and technology on various industries; Chapter 4 analyzes the regional effects of nuclear science and technology; Chapter 5 investigates human factors related to nuclear science and technology; Chapter 6 addresses environmental and societal demands for nuclear science and technology; Chapter 7 discusses nuclear science and technology in the context of human development; and Chapter 8 outlines the future of humanity and nuclear civilization.

Chapter 1 was authored by a team of five, which includes Tao Zhou, Yao Yao, and Huaichang Lu; Chapter 2 was authored by a team of six, which includes Yao Yao, Tao Zhou, and Wenbin Liu; Chapter 3 was authored by a team of eight, which includes Tao Zhou, Yao Yao, and Chunmei Liu; Chapter 4 was authored by a team of eight, which includes Yao Yao, Tao Zhou, and Dongli Huang; Chapter 5 was authored by a team of five, which includes Shang Mao, Tao Zhou, and Jianyu Tang; Chapter 6 was authored by a team of five, which includes Tao Zhou, Shang Mao, and Dong Wei; Chapter 7 was authored by a team of five, which includes Shang Mao, Tao Zhou, and Chunhui Xue; Chapter 8 was authored by a team of five, which includes Peng Xu, Shang Mao, and Tao Zhou. The entire book was compiled by Tao Zhou, Yao Yao, and Shang Mao. In addition, Xiaofang Liu, Peng Liu, Tianyu Gao, Shilei Dun, Zongyan Bai, Shuyu He, Bokang Chang, Biao Zhang, and Da Chen from Southeast University; Lanyu Zhou from China Nuclear Power Engineering Co., Ltd.; and the faculty and students of the Department of Nuclear Science and Technology, School of Energy and Environment, Southeast University, provided substantial assistance in material retrieval, text entry, and illustration preparation. Special acknowledgment is due to Shang Mao from Zhengzhou University of Light Industry for his significant contributions to the overall structure. We express our sincere gratitude to all of them.

Tao Zhou
Department of Nuclear Science and Technology,
Institute of Nuclear Thermal-hydraulic Safety and Standardization,
Southeast University, Nanjing, Jiangsu, China

Yao Yao
Department of Nuclear Science and Technology,
Institute of Nuclear Thermal-hydraulic Safety and Standardization,
Southeast University, Nanjing, Jiangsu, China

Shang Mao
School of Energy and Power Engineering,
Zhengzhou University of Light Industry,
Zhengzhou, Henan, China

Chapter 1

Definition, function, and classification of nuclear science and technology

Tao Zhou[a], Yao Yao[a], Huaichang Lu[a], Peng Xu[b] and Xiaofang Liu[a]

[a] *Department of Nuclear Science and Technology, Institute of Nuclear Thermal-Hydraulic Safety and Standardization, Southeast University, Nanjing, Jiangsu, China,* [b] *Institute of Advanced Nuclear Energy Technology, State Power Investment Corporation Research Institute, Beijing, Beijing, China*

1.1 Nuclear science

1.1.1 Definition of nuclear science

Nuclear science is the discipline that studies the structure, properties, and reactions of atomic nuclei, as well as the applications of nuclear energy. Based on fundamental theories such as quantum mechanics and relativity, it explores fundamental phenomena including nuclear stability, radioactive decay, and nuclear reactions.

1.1.2 Role of nuclear science

1.1.2.1 Scientific research

Nuclear science provides new tools and methods for basic scientific research, promoting the development of related disciplines such as physics, chemistry, and biology. Concurrently, research findings in nuclear science offer crucial theoretical support and technical means for the advancement of other disciplines.

1.1.2.2 Social development

The role of nuclear science encompasses multiple aspects, including energy, medicine, industry, agriculture, and national security, holding significant value for the development and progress of modern society.

Nuclear Science, Technology, and Civilization. DOI: https://doi.org/10.1016/B978-0-443-13941-3.00001-X
Copyright © 2026 Elsevier Inc. All rights are reserved, including those for text and data mining, AI training, and similar technologies.

1.1.2.3 National Security

Nuclear science plays a critical role in national security, particularly in the development and maintenance of nuclear weapons. The development and control of nuclear weapons, along with the formulation and implementation of international nuclear nonproliferation treaties, are of great value to global security.

1.1.3 Classification within the field of nuclear science

1.1.3.1 Nuclear physics

Although born just over a century ago, nuclear physics has developed into a discipline with substantial impact on politics, military affairs, economy, science, and technology and is inextricably linked to human life. Nuclear physics, also known as "atomic nucleus physics," is an important branch of physics. It primarily studies various subatomic particles and their relationships, as well as nuclear energy, including the structure, properties, and reactions of atomic nuclei.

1.1.3.2 Particle physics

Particle physics, also known as "high-energy physics," is a branch of physics. It studies the structure and properties of matter in the microscopic world, at a level deeper than the atomic nucleus, and the phenomena of mutual transformation of this matter at very high energies, along with the causes and laws that govern these phenomena. As research deepens the content of particle physics continues to evolve. Particle physics is at the forefront of contemporary physics development and will alter humanity's fundamental understanding of nature. Research in particle physics challenges human preconceptions, stimulating and driving human knowledge toward its most basic levels. Particle physics is experimentally based, yet it relies heavily on the close integration of experiment and theory. The study of the microscopic structure of matter conducted in particle physics is fundamental to research across various disciplines. Achievements and every advancement in microscopic matter research have significant applications in physical chemistry, materials science, biology, medicine, and agriculture. In-depth experimental research in particle physics relies on high-energy accelerators; these accelerators often require the application of numerous advanced technologies, such as superconductivity, precision mechanics, automatic control, and computer technology. These technologies often, within a relatively short period, greatly propel the high-tech development of the entire society.

1.1.3.3 Radiochemistry

Radiochemistry studies the chemical properties and applications of radioactive isotopes. Radioisotopes are isotopes with unstable atomic nuclei that spontaneously emit radiation. As one of the sources of nuclear technology applications, their use permeates national defense, industry, agriculture, medicine, scientific research, and other fields. Their production and supply directly impact national security, economic development, and the improvement of people's health. Classified by application, radioactive isotopes include medical isotopes, industrial isotopes, and research isotopes.

1.1.3.4 Nuclear engineering

Nuclear engineering involves the generation, conversion, and application of nuclear energy[1]. It includes the design and operation of nuclear power plants and also encompasses key scientific and technological areas such as the nuclear fuel cycle, radiation protection, and nuclear waste management. It is dedicated to achieving the safe, efficient, and sustainable development of nuclear energy.

1.1.3.5 Nuclear medicine

Nuclear medicine is a comprehensive interdisciplinary field that utilizes nuclides and nuclear technology for fundamental life sciences and for clinical disease diagnosis and treatment. It primarily consists of two branches: clinical nuclear medicine and basic nuclear medicine. Basic nuclear medicine mainly uses the properties of nuclides and radiation for fundamental research in life sciences and basic medicine. Clinical nuclear medicine is functionally divided into diagnostic nuclear medicine and therapeutic nuclear medicine and also includes nuclear medicine equipment, associated pharmaceuticals, and in vitro diagnostic reagents. As a product of the integration of nuclear technology with medicine and related disciplines, nuclear medicine is both an important field of nuclear technology applications and a vital component of modern medicine.

1.1.3.6 Astrophysics

Astrophysics is not only an important part of astronomy but also a significant branch of physics. This discipline utilizes physical theories, methods, and techniques to research and explore the evolutionary laws, chemical composition, physical conditions, structure, and morphology of celestial bodies. Astrophysics is mainly divided into cosmogony, cosmochemistry, cosmology, galactic astronomy, planetary physics, stellar astronomy, solar system physics, stellar physics, etc. Furthermore high-energy astrophysics, space astronomy, radio astronomy, etc. are also branches of this discipline. By analyzing the electromagnetic radiation of celestial bodies using physical methods and techniques, various physical parameters of celestial bodies can be obtained. Combining the obtained parameters with physical theories to explain the physical processes and evolutionary processes occurring on celestial bodies is a crucial task of theoretical astrophysics and observational astrophysics. It also involves studying nuclear processes in the universe, such as energy generation in stars and supernova explosions.

1.2 Nuclear technology

1.2.1 Definition of nuclear technology

Nuclear technology refers to applied technologies based on the principles of nuclear science. It encompasses techniques for research, development, and application utilizing the physical properties and reaction processes of atomic nuclei. This includes the generation of nuclear energy, the preparation and application of radioactive isotopes, and nuclear radiation technology.

1.2.2 Role of nuclear technology

1.2.2.1 Industrial applications

Applications of nuclear technology in the industrial sector include nondestructive testing, material modification, and radiation processing. These technologies play a significant role in improving product quality, ensuring safe production, and promoting the development of new materials.

1.2.2.2 Agricultural development

Applications of nuclear technology in agriculture include crop breeding, radiation processing of agricultural products and food, and pest control. These technologies help enhance agricultural productivity, improve the quality of agricultural products, and ensure food safety.

1.2.2.3 Medical health

Nuclear technology has extensive applications in medicine. For example, radioactive isotopes and ionizing radiation are used for diagnosis and treatment. Radiotherapy, proton therapy, and heavy-ion therapy are effective methods for treating cancer using nuclear technology.

1.2.2.4 Environmental protection

Applications of nuclear technology in environmental protection include air pollutant monitoring, water pollution control, and soil remediation. These technologies contribute to enhancing environmental monitoring capabilities, improving environmental quality, and protecting ecosystems.

1.2.2.5 Public safety

In the field of public safety, security screening during travel is one application of nuclear technology. For instance, through CT security scanners and human X-ray inspection systems, security personnel gain "perspective" capabilities, enabling rapid and accurate identification of dangerous goods, thus ensuring travel safety.

1.2.2.6 Other fields

Nuclear technology can also be used in oil well logging, haze source apportionment, archaeology, forensics, and other fields. In addition, nuclear energy can power space exploration, expanding the scope of human activity. It has become one of the driving forces for continuous innovation and development in new technologies, new materials, new processes, and new methods.

1.2.3 Classification within the field of nuclear technology

1.2.3.1 Nuclear power technology

Design, construction, and operation of nuclear power plants; mining, processing, and reprocessing of nuclear fuel [2].

1.2.3.2 Radioisotope technology

Preparation and labeling of radioactive isotopes; applications in medicine, industry, and scientific research.

1.2.3.3 Radiation technology

Preparation and application of radiation sources; radiation protection and safety measures.

1.2.3.4 Nuclear agricultural technology

Radiation mutation breeding; food preservation by irradiation.

1.2.3.5 Nuclear medicine technology

Development and practical application of radiopharmaceuticals and radio diagnostic and radiotherapeutic techniques.

1.3 Nuclear energy

1.3.1 Definition of nuclear energy

1.3.1.1 Basic definition

Nuclear energy (or atomic energy) is the energy released from atomic nuclei via nuclear reactions. In the early 20th century, physicists discovered a more efficient way to release energy—nuclear reactions. Unlike chemical reactions, during a nuclear reaction, the element type of the atom itself changes. The mass of the matter produced after a nuclear reaction is less than the mass before the reaction; the reduced portion of matter is converted into energy. This energy is called nuclear energy. The nuclear force is a short-range force capable of generating strong attraction over very short distances. The nature of the nuclear force determines the stability of atomic nuclei. Heavier atomic nuclei can release energy through fission; lighter atomic nuclei can release energy through fusion.

1.3.1.2 Definition of nuclear fission

Nuclear fission refers to a form of nuclear reaction in which a heavy atomic nucleus (primarily uranium or plutonium nuclei) splits into two or more nuclei of smaller mass. The energy released in fission relates to the way mass-energy is stored in the nucleus. From the heaviest elements down to iron, the efficiency of energy storage is essentially continuously variable. Therefore any process by which a heavy nucleus can split into lighter nuclei (down to iron) is energetically favorable. If the nucleus of a heavier element can split and form lighter nuclei, energy is released.

1.3.1.3 Definition of nuclear fusion

Nuclear fusion, also known as "fusion reaction or thermonuclear reaction," is the process where light atomic nuclei (e.g., deuterium and tritium) combine to form a heavier atomic nucleus (e.g., helium), releasing enormous energy. Because chemistry is the science that studies the properties, composition, structure, and laws of change of matter at the molecular and atomic level, whereas nuclear fusion occurs at the nuclear level, nuclear fusion is not considered a chemical change.

1.3.1.4 Definition of nuclear decay

Nuclear decay is the process by which an atomic nucleus spontaneously emits a particle and transforms into another nucleus. This includes the process where a nucleus captures an electron from the K or L shell, decaying into another atomic nucleus with an atomic number reduced by one and the same mass number. As the K shell is closest to the nucleus, K-capture occurs most readily. When K-capture occurs, an outer electron invariably fills the inner shell vacancy, emitting characteristic X-rays specific to the daughter element. This energy may also be transferred to a more outer electron, causing it to be emitted as a free electron. This electron is called an "Auger electron," resulting from nuclear decay.

1.3.2 Role of nuclear energy

1.3.2.1 Energy production

Nuclear power plants utilize nuclear fission to generate energy, representing an efficient and clean form of energy. As a clean, low-carbon, safe, and efficient new energy source, nuclear energy represents an important direction for future energy development. Nuclear power generation is already widely used worldwide, playing a significant role in ensuring national energy security, optimizing energy structures, improving air quality, promoting green development, and serving the

Belt and Road Initiative [3]. In addition, nuclear energy can be used for multipurpose applications such as district heating, environmental improvement, powering artificial intelligence, and providing process heat for hydrogen production.

1.3.2.2 Nuclear power

Nuclear power can be applied in nuclear-powered aircraft carriers, nuclear submarines, and nuclear-powered icebreakers. In the future, it may provide power sources for nuclear-powered automobiles, trucks, cargo ships, etc.

1.3.2.3 Space and deep-sea exploration

Radioisotope thermoelectric generators provide energy for space and deep-sea probes.

1.3.3 Classification within the field of nuclear energy

1.3.3.1 Nuclear fission field

Nuclear power plants and atomic bombs represent the two main application areas of nuclear fission energy[4]. The primary mechanistic difference lies in whether the chain reaction rate is controlled. The key equipment in a nuclear power plant is the nuclear reactor, analogous to the boiler in a thermal power plant; it is where the controlled chain reaction occurs. Nuclear fission [5] is also indispensable for important fundamental questions, such as the synthesis of superheavy new elements, the evolution of elements in the universe, and the study of reactor neutrinos.

1.3.3.2 Nuclear fusion field

Humanity has already achieved uncontrolled nuclear fusion, as demonstrated in hydrogen bomb explosions. Controlled nuclear fusion can be used for electricity generation, heating, and other fields. Furthermore controlled fusion engines hold the potential to significantly enhance the shipping capacity of large vessels such as nuclear-powered aircraft carriers and submarines. In addition, powering artificial intelligence represents a new pathway for future fusion development.

1.3.3.3 Nuclear decay field

The radioactivity produced by nuclear decay has important applications in research across many disciplines. For example, the study of alpha decay energy spectra provides information on nuclear structure and can be used in archaeology. Another example is the use of the natural radioactivity of carbon-14 to determine the age of biological remains or measuring the content of two elements in a rock to determine its geological age.

1.4 Nuclear safety culture

1.4.1 Definition of nuclear safety culture

1.4.1.1 Basic definition

The term "nuclear safety culture" was first introduced by the International Nuclear Safety Advisory Group (INSAG) of the International Atomic Energy Agency (IAEA) in its 1986 "Summary Report on the Post-Accident Review Meeting on the Chernobyl Accident." In 1988 INSAG further elaborated on safety culture as a fundamental management principle for nuclear power safety, emphasizing that the achievement of safety goals must permeate all activities within a nuclear power plant. In 1991 INSAG provided a clear definition of nuclear safety culture in "Safety Culture" (INSAG-4)[6], defining it as the assembly of characteristics and attitudes in organizations and individuals that establishes that nuclear plant safety issues receive the attention warranted by their significance.

1.4.1.2 Connotation of nuclear safety

Nuclear safety in a broad sense refers to the achievement of proper operating conditions, prevention of accidents, or mitigation of accident consequences, resulting in protection of workers, the public, and the environment from undue radiation hazards from nuclear facilities, nuclear activities, nuclear materials, and radioactive substances by taking necessary monitoring, protection, prevention, and mitigation measures. Nuclear safety in a narrow sense focuses specifically on the technical and organizational measures taken during the design, construction, operation, and decommissioning of nuclear facilities to protect personnel, society, and the environment from potential radiation hazards.

1.4.1.3 Connotation of nuclear safety culture

1.4.1.3.1 Cornerstone of nuclear safety for organizations and individuals

Safety culture involves the behaviors and attitudes of both organizations and individuals, emphasizing the importance of correct understanding and appropriate action in nuclear safety issues. Safety culture is not only related to individual work attitudes and thinking habits but is also closely linked to the work style of the organization.

1.4.1.3.2 Manifestation from abstract to concrete

Although intrinsic qualities such as work attitudes and thinking habits are relatively abstract, these qualities can be tested and verified through concrete manifestations to ensure their alignment with nuclear safety objectives.

1.4.1.3.3 Demands for high vigilance and sense of responsibility

Safety culture requires that personnel, when performing safety-related duties, possess high vigilance, accurate judgment, ample knowledge, and demonstrate a strong sense of responsibility to ensure the comprehensive implementation of nuclear safety.

1.4.2 Role of nuclear safety culture

1.4.2.1 Ensuring safe operation of the nuclear industry

Nuclear safety culture is the cornerstone of safe operation in the nuclear industry. It requires organizations and individuals to consistently prioritize safety in all activities. In the nuclear field, safety is not merely a technical issue but also a cultural phenomenon. By establishing a safety-first mindset and behavioral habits, accidents can be effectively prevented, ensuring the safety of nuclear facilities, nuclear activities, nuclear materials, and radioactive substances, thereby protecting the environment and public health.

1.4.2.2 Enhancing organizational management level

The essence of nuclear safety culture lies in establishing a set of scientific and rigorous regulations and organizational systems. It prompts organizations to formulate comprehensive and clear safety policies, develop and implement detailed safe operating procedures, and ensure all activities comply with safety standards. However, mechanical execution of sound procedures and good working methods is insufficient; nuclear safety culture demands that all duties important to safety be performed correctly. This contributes to improving the overall management level of the organization.

1.4.2.3 Promoting cultural improvement

Nuclear safety culture emphasizes continuous improvement, constantly seeking opportunities for enhancement, and improving safety performance through learning and innovation. Organizations conduct regular safety audits to assess the effectiveness of their safety management systems and identify areas for improvement. Establishing continuous improvement mechanisms, encouraging employees to propose improvement suggestions, and adjusting based on evaluation results enables the organization to adapt to technological developments and environmental changes, continuously raising the level of safety management.

1.4.2.4 Internalizing safety concepts, enhancing employee safety awareness and sense of responsibility

Nuclear safety culture needs to be internalized in the heart and externalized in action, advocating that the nuclear safety philosophy prioritizing safety above all else becomes the conscious action of all personnel. It links each individual's work attitude and thinking habits with the organization's work style, requiring employees to perform all safety-important duties with high vigilance, real-time insight, abundant knowledge, accurate judgment, and a high sense of responsibility, thereby enhancing employee safety awareness.

1.4.2.5 Motivating employee participation

By establishing safety incentive mechanisms that reward employees or teams with outstanding safety performance, all employees can be motivated to place greater emphasis on safety. Simultaneously, encouraging employees to actively participate in safety affairs, provide feedback, and play a role in safety improvements further enhances their sense of responsibility.

1.4.2.6 Promoting knowledge sharing

Nuclear safety culture encourages employees to share safety knowledge and experiences, promoting the internal dissemination of safety culture through case studies, seminars, and other forms. Furthermore organizations learn from the safety management best practices of other entities within the industry, continuously assimilating new concepts and technologies to enhance their own safety culture level.

1.4.2.7 Enhancing social trust

The construction of a nuclear safety culture pertains not only to the safety of the nuclear industry itself but also reflects responsibility toward public safety and the environment. By establishing a robust nuclear safety culture, organizations can better respond to various challenges and ensure the safe operation of nuclear facilities. Timely and accurate communication of safety information to the public [7] enhances societal trust in the nuclear industry, creating favorable conditions for its sustainable development.

1.4.3 Approaches to building nuclear safety culture

1.4.3.1 Strengthening nuclear safety awareness education

Enhancing public understanding of nuclear safety issues and increasing nuclear safety awareness through various channels and formats.

1.4.3.2 Improving laws and regulations

Establishing and refining the relevant legal and regulatory framework for nuclear safety, clearly defining management responsibilities and requirements.

1.4.3.3 *Strengthening technical and managerial measures*

Improving the safety design and construction standards of nuclear power facilities; strengthening the management and regulation of nuclear power facilities [8].

1.4.3.4 *Promoting cross-fertilization with new productive forces*

Actively promoting cross-cutting technologies with new productive forces, including artificial intelligence, to foster the development of an intelligent safety culture.

1.4.3.5 *Enhancing international cooperation*

Collaborating to address nuclear safety challenges, sharing advanced experiences and practices.

1.5 Human civilization

1.5.1 Definition and role of human civilization

1.5.1.1 *Definition*

Human civilization refers to the totality of the material wealth, spiritual wealth, and corresponding social systems, culture, arts, science, and technology created by humanity over the course of its long historical development. It embodies humanity's progress and accumulation in areas such as social organization, language and writing, morality and ethics, and legal systems.

1.5.1.2 *Function*

Human civilization not only carries historical memory and cultural heritage but also drives technological innovation and social transformation. It reflects humanity's exploration and understanding of the living environment, the natural world, and its own existence. It represents the collective achievement of humankind in its pursuit of progress, harmony, and sustainable development. Nuclear science and technology, as major breakthroughs of the 20th century, have had a profound impact on human civilization. Driven by artificial intelligence the strong revitalization of nuclear science and technology has arrived. All developmental efforts point toward a common direction—building an energy foundation capable of supporting a future intelligent civilization.

1.5.2 Impact of nuclear science on the development of human civilization

1.5.2.1 *Scientific revolution*

The rise of nuclear science [9,10] marked an important milestone in modern physics, catalyzing a revolution in scientific methods and theories.

1.5.2.2 *Transformation of energy concepts*

As an efficient and clean energy source, nuclear energy has changed people's perspectives on and usage of energy.

1.5.2.3 *Ethical and safety issues*

The development of nuclear science has also brought about ethical and safety concerns, such as the threat of nuclear weapons and the risks of nuclear accidents.

1.5.3 Impact of nuclear technology on the development of human civilization

1.5.3.1 *Energy production*

The application of nuclear technology [11,12] in energy production provides new pathways for addressing the energy crisis and reducing greenhouse gas emissions.

1.5.3.2 *Industry and scientific research*

Applications of nuclear technology in industry and research [13–17], such as nondestructive testing, materials science, and fundamental physics research, have driven technological progress in industry and research-related fields.

1.5.3.3 *Agriculture and food*

Applications of nuclear technology in agriculture and food, such as radiation breeding, irradiation storage, and food safety control research, have enhanced food security and food quality, promoting technological advancement in agriculture and food-related fields.

1.5.3.4 *Medicine and health*

The application of radioactive isotopes in medicine, including cancer treatment and medical imaging examinations, has greatly improved medical standards and contributed to the enhancement of public health.

1.5.3.5 *Military and international politics*

The application of nuclear technology in the military domain, especially the existence of nuclear weapons, has had a significant impact on the international political landscape.

1.5.3.6 *Environment and society*

The environmental impact of nuclear technology [18–20], including nuclear waste management and the risk of nuclear accidents, has also drawn societal attention to sustainable development.

References

[1] X.J. Tang, C. Zhang, Application research of PSA risk analysis method in nuclear power plants, Electr. Eng. Technol. 2019 (20) (2019) 102–104.

[2] T. Zhou, H.L Zhang, Y. Yao, et al., Development trends of information technology and advanced nuclear power, South. Energy Constr. 12(4)(2025) 109–114.

[3] S. Yang, Study on the Influence of Steam Bypass Discharge on the Response Characteristics of the Primary Loop Under Small-Break LOCA, North China Electric Power University, Beijing, 2023.

[4] M2 PressWIRENuclear Safeguards: AUKUS Statement to the IAEA Board of Governors, M2 PressWIRE (The UK), 2024.

[5] Y. Lu, Research on the Three Mile Island Nuclear Accident From the Perspective of Technological Explanation, Donghua University (China), 2021.

[6] P.L Hu, H.F Chen, L. Yuan, et al., Problems and lessons of emergency evacuation in three major nuclear accidents, Chin. J. Radiol. Health (2024) 1–10. http://kns.cnki.net/kcms/detail/37.1206.R.20241204.1654.002.html.

[7] T. Fujita, Energy knowledge and public response to restarting nuclear plants in Japan following the Fukushima accident, Util. Policy 92 (2025) 101858.

[8] K. Mehboob, C. Xinrong, R. Ahmed, US NRC progress in source term evaluation, Int. J. Nucl. Energy Sci. Technol. 8 (1) (2014) 12–20.

[9] L. Pei, S.W. Zhou, Risk analysis and mitigation measures for station blackout accidents in nuclear power plants, Nucl. Power Eng. 39 (6) (2018) 74–78.

[10] F.J. Wang, Study on steam generator tube rupture accident handling in nuclear power plants, Manage. Technol. Small Medium Enterp. (Late Ed.) 2020 04 (2020) 156–157.

[11] W.T. Cai, Steady-State Operating Characteristics and Sensitivity Analysis of Main Steam Line Break Accident in Small PWRs, Shenzhen University (China), 2018.

[12] M.C. Zhang, Study on the Most Likely Rupture Locations of Main Steam and Feedwater Pipelines in Tianwan Nuclear Power Plant, Harbin Institute of Technology, 2014.

[13] J.Z. Zhu, Reactor Safety Analysis, Xi'an Jiaotong University Press (China), 2004.

[14] National Nuclear Safety Administration (China)Deterministic Safety Analysis of Nuclear Power Plants, 2021.

[15] M.Z. Choudri, Integrated Deterministic and Probabilistic Safety Analysis of the AP1000 Passive Containment Cooling System, North China Electric Power University (China), 2019.

[16] X. Ran, X.H. Zhang, J. Li, F. Yang, P. Wu, Review of best estimate plus uncertainty (BEPU) methods in nuclear power plants, Sci. Technol. Vision (24) (2015) 11–13 16.

[17] Z.L. Wang, Study on the Motion of Solid-Liquid Fine Particles in Narrow Channels, North China Electric Power, University, China, 2013.

[18] H. Ding, Development of the Uncertainty and the Sensitivity Analysis Function Module in the Nuclear Power Station PSA, Software, Hefei University of Technology, 2005.

[19] J.W. Li, Design of containment performance test in nuclear power plants, Electromech. Inf. 2011 (6) (2011) 20–22.

[20] D.Y. Mao, D.Z. Xue, Bayesian methods for reliability data processing in PSA analysis of nuclear power plants, Nucl. Power Eng. (5) (2000) 451–455.

Chapter 2

History and development of nuclear science and technology

Yao Yao, Tao Zhou, Wenbin Liu, Tianyu Gao, Haolei Zhang and Shilei Dun

Department of Nuclear Science and Technology, Institute of Nuclear Thermal-hydraulic Safety and Standardization, Southeast University, Nanjing, Jiangsu, China

2.1 Origin of nuclear science and technology

2.1.1 Discovery of X-rays

2.1.1.1 Phenomenon discovery

The late 19th century witnessed profound transformations in physics. Advances in electromagnetic theory and innovations in experimental techniques created fertile ground for new discoveries. On November 8, 1895, German physicist Wilhelm Conrad Röntgen [1,2] accidentally discovered X-rays while studying cathode rays.

2.1.1.2 Experimental verification

During his cathode ray experiments, Röntgen [3] observed that chemicals on a fluorescent screen emitted light even when no electron beam directly struck it. This phenomenon sparked his curiosity. Through a series of experiments, he demonstrated that this new type of radiation possessed extraordinary penetrating power—capable of passing through books, wood, and even human skin to form images on the fluorescent screen.

2.1.2 Discovery of radioactive elements

2.1.2.1 Antoine Henri Becquerel discovers radioactivity

While studying uranium salts in 1896, Antoine Henri Becquerel uncovered a novel natural phenomenon—radioactivity [4]. He observed that these salts spontaneously emitted rays, marking humanity's first recognition of atomic nucleus instability.

Nuclear Science, Technology, and Civilization. DOI: https://doi.org/10.1016/B978-0-443-13941-3.00002-1
Copyright © 2026 Elsevier Inc. All rights are reserved, including those for text and data mining, AI training, and similar technologies.

2.1.2.2 The Curies discover polonium and radium

Becquerel's discovery sparked immense interest in the Curies, who began investigating these mysterious rays. In 1898 they isolated two new radioactive elements from pitchblende ore: polonium and radium [5]. The Curies [6,7] processed vast quantities of pitchblende ore, employing complex chemical processes to isolate and purify the radioactive elements. Working under extremely primitive laboratory conditions, they succeeded after countless experiments and attempts in extracting minute quantities of radium from several tons of ore.

2.1.3 Types of radiation and atomic nucleus models

2.1.3.1 Distinguishing types of radiation

In the early 20th century, research into the nature of radiation became a hot topic in physics. In 1899 Rutherford [8] successfully distinguished alpha rays from beta rays through a series of carefully designed experiments. He discovered that alpha rays consist of positively charged particles—twice the mass of a helium nucleus—possessing strong ionizing power but weak penetrating ability. In contrast, β rays comprised negatively charged electrons, exhibiting high penetrating power but weaker ionizing ability. Rutherford's experiments not only revealed the physical properties of rays but also provided crucial clues for understanding the internal structure of atoms.

2.1.3.2 The proposal of the atomic nucleus model

In 1911 Rutherford proposed the atomic nucleus model, positing that atoms consist of a dense positively charged core with electrons orbiting around it. This model was based on the results of alpha particle scattering experiments, where a very small fraction of alpha particles underwent large-angle scattering upon passing through gold foil. This phenomenon indicated the existence of an extremely small yet massively dense core within the atom—the atomic nucleus.

2.2 Origin and development of nuclear energy

2.2.1 Discovery of the atom

2.2.1.1 Classical atomic theory

2.2.1.1.1 Foundational origins

Ancient philosophers engaged in profound contemplation regarding the fundamental building blocks of matter. Ancient Greek atomism represented a natural philosophical conception of matter and its composition, which was lost during the Roman Empire [9,10]. It was not until the mid-15th century Renaissance that it was unexpectedly rediscovered, gaining prominence and revival. It became a cornerstone of the material conception in the modern scientific revolution, forming the basis for the mechanical systems developed by Galileo and Newton. Ancient Greek atomism is represented by the theories of Democritus and Epicurus.

2.2.1.1.2 Democritus' atomism

Democritus' atomism posited that the fundamental constituents of all things are atoms and void, with atoms being the smallest indivisible entities. All matter is composed of atoms and void, and atoms are indivisible and unchangeable. Atoms possess hooks and horns, interlocking mechanically to form structures. He posited that natural motion in the cosmos occurs through vortex motion, where atoms classify and collide; interlocking via hooks and horns, they coalesce into objects and celestial bodies (Democritus, Ethica). Thus Marx described Democritus' atomism as deterministic and mechanistic.

2.2.1.1.3 Epicurus' atomism

Epicurus' atomism developed from Democritus' theory, asserting that atoms possess internal structure yet are indivisible—for division would lead to infinite division and dissolution into nothingness. He further posited that atoms naturally move in straight lines through the void toward the cosmic center. During parallel motion, accidental deviations cause collisions, ultimately generating vortex motion, classification, and combination. He posited that atomic combinations were not mechanical interlocking but rather seemed to arise when atoms, in their vertical descent, deviated because of self-awareness, attracting, combining, and repelling each other to form all things.

2.2.1.2 Modern atomic theory

True scientific inquiry began in the 17th century. With the rise of the scientific method, scientists started verifying atomic concepts through experiments. Among them, British chemist John Dalton proposed atomic theory in 1808, asserting that matter consists of minute ultimate particles called atoms; atoms are indivisible and cannot be created or destroyed; atoms of a given element are identical and possess a constant atomic weight; different elements have distinct atomic weights; the weight of a compound particle equals the sum of the atomic weights of its constituent elements; for known elements with more than one compound, the number of atoms of each element in these compounds is proportional. This theory laid the foundation for subsequent research [11].

2.2.1.3 Developmental connections

2.2.1.3.1 Philosophical origins of atomic theory and early scientific development

Throughout the history of science the atom has remained a core9 concept. From ancient philosophical contemplation to modern precise experimentation, humanity's exploration of atoms has never ceased. From Democritus to Newton, atomic theory primarily belonged to the realm of natural philosophy, where atoms were viewed as a philosophical construct to explain the cosmos. Newton introduced the atomic concept into mechanics, regarding atoms as carriers of force—the smallest particles or material points—thus establishing the foundation for Newtonian theory and theoretical physics. In modern physics history the ether wave theory faced four waves of opposition from atomist schools, gradually fading into obscurity. Newton's critique of the ether wave theory in 1684 led to the particle theory dominating for over a century.

2.2.1.3.2 The rise of field theory and the atom-energy debate

In 1844 Faraday abandoned the ether theory, proposing magnetic lines of force and field theory. In 1895 the atomic theory school represented by Boltzmann engaged in a debate with the energy-only theory school represented by Ostwald, resulting in the defeat of the energy-only theory.

2.2.1.3.3 Experimental breakthroughs and the discovery of wave-particle duality

The framework of the ether theory began to crumble in the 1880s when British chemist William Crookes discovered cathode rays, composed of negatively charged particles, through his experiments. In 1905 Einstein disproved the existence of the ether and proposed the special theory of relativity, which received experimental support from atomic structure studies, ultimately establishing wave-particle duality.

2.2.1.3.4 The outcome of debates and the establishment of modern atomic theory

The debates between the Copenhagen school (represented by Bohr) and Einstein further propelled scientific progress. The controversies surrounding atomic theory versus ether wave theory and monism versus field theory profoundly influenced the course of science. The fundamental distinction between modern atomic theory and classical atomic theory lies in recognizing atoms as having internal composition and being divisible. This shattered the framework of assumptions about atoms being indivisible and unchangeable, which had persisted since Democritus.

2.2.2 The discovery of nuclear energy

2.2.2.1 The connection between nuclear energy and radioactivity

2.2.2.1.1 The discovery of radioactivity and the energy stored in the atomic nucleus

In 1895 German physicist Wilhelm Röntgen at the University of Würzburg accidentally discovered X-rays during an experiment, sparking a global scientific revolution. Inspired by this, French scientist Henri Becquerel further discovered in 1896 that uranium salts spontaneously emitted radiation, revealing natural radioactivity for the first time. This discovery provided humanity's initial understanding that immense energy resides within atomic nuclei, laying the scientific foundation for nuclear energy development and application [12]. It marked humanity's entry into a new era of exploring and harnessing nuclear energy.

2.2.2.1.2 Discovery of elemental decay and elucidation of atomic structure

In 1898 Ernest Rutherford discovered the radiation emitted by thorium, composed of three types of rays: alpha, beta, and gamma. Between 1900 and 1904, he further determined that alpha rays consisted of helium nuclei, beta rays were streams of electrons, and he and later the French scientist Villard discovered gamma rays, proving them to be electromagnetic waves. In 1903 Rutherford and Soddy mapped the decay sequences of radioactive element families, thereby demonstrating that inorganic substances also follow objective laws of evolution. This marked the first historical discovery of an element spontaneously decaying into its isotopes and new elements. The discovery of radioactive radiation and elemental decay provided favorable conditions for understanding atomic structure, radioactive decay processes, and unraveling the microscopic composition within atoms.

2.2.2.2 *The discovery of nuclear fission*

2.2.2.2.1 Theoretical foundations of nuclear energy and its existence

The theoretical foundation of nuclear energy lies in Einstein's theory of relativity, expressed by the famous equation $E = mc^2$. This equation demonstrates the equivalence of mass and energy, with the proportionality constant being the square of the speed of light. Einstein's theory was published in 1905, and since then, nations worldwide have conducted multifaceted explorations. Through laboratory observations and analyses, scientists confirmed the theory's validity and the existence of nuclear energy. Particularly intriguing to nuclear experts was the phenomenon where a uranium nucleus, after absorbing a neutron, splits into two parts.

2.2.2.2.2 Energy estimation and controlled chain reaction experiments

In February 1939 Meitner and Frisch first demonstrated that uranium nuclei split like droplets, coining the term "fission" to describe nuclear fission. More significantly, based on Bohr's liquid-drop model, they estimated that a single nuclear fission releases approximately 200 MeV of energy. In the United States the significance of nuclear fission was quickly recognized. In 1942, the University of Chicago laboratory successfully achieved a controlled chain reaction using ^{235}U, providing conclusive evidence for atomic bomb development and nuclear energy utilization. In 1946 Chinese scientists Qian Sanqiang and He Zehui, working in France, discovered triple and quadruple fission phenomena in uranium, offering new evidence for subsequent nuclear reaction theories.

2.2.2.3 The discovery of nuclear fusion

2.2.2.3.1 Nuclear fusion process and reaction characteristics

Nuclear fusion, also known as "fusion reaction" or "fusion process," refers to a nuclear reaction where two lighter nuclei combine to form a heavier nucleus and an extremely light nucleus (or particle). In this process, matter is not conserved because part of the mass from the fusing nuclei is converted into photons (energy). During fusion the two lighter nuclei experience a mass deficit and release enormous energy. Although both nuclei carry positive charges and repel each other, when two sufficiently energetic nuclei collide head-on, they can come together closely enough for the nuclear force to overcome the Coulomb repulsion, triggering the nuclear reaction known as "fusion."

2.2.2.3.2 Discovery and mechanism research of nuclear fusion

Nuclear fusion was discovered in 1932 by Australian scientist Mark Oliphant. Through experiments, he observed hydrogen isotopes (such as deuterium and tritium) combining under specific conditions to form helium nuclei while releasing energy. This discovery laid the foundation for nuclear fusion research. In his experiments, Oliphant employed ultrahigh temperatures and pressures to induce hydrogen nuclei to fuse into helium nuclei while emitting neutrons. This process resembles the energy generation mechanism within stars, though sustaining stable reactions under laboratory conditions proved challenging. Subsequently, in the early 1950s, Mark Oliphant established the Australian Plasma Fusion Research Facility at the Australian National University, initiating research into nuclear fusion.

2.2.3 Development of nuclear energy

2.2.3.1 The dawn of the nuclear age

2.2.3.1.1 The discovery of heavy nuclear fission and chain reactions

For historical reasons, nuclear energy was initially developed and utilized primarily as a devastating weapon of mass destruction. In the late 1930s the discovery of heavy nuclear fission and chain reactions[13] laid the scientific foundation for harnessing nuclear energy.

2.2.3.1.2 The Manhattan Project's atomic bomb development plan

To prevent Nazi Germany from exploiting this scientific breakthrough to create nuclear weapons and enslave humanity the US government, urged by the scientific community, passed a resolution in December 1941 to develop an atomic bomb. In August 1942 the Manhattan Project—a top-secret initiative to build an atomic bomb—was formally launched.

2.2.3.1.3 Construction of the nuclear reactor at the University of Chicago

To demonstrate the practical conditions for achieving a chain reaction the United States decided to build a self-sustaining nuclear reactor. Under Fermi's leadership, construction began at the University of Chicago in December 1941. The reactor used natural uranium as fuel, graphite as a moderator, and cadmium rods that could be inserted and withdrawn at any time as control rods. On December 2, 1942 this reactor was completed and commenced operation, marking the first sustained chain reaction in history and opening the chapter of human utilization of nuclear energy.

2.2.3.1.4 Formation of the military nuclear industrial complex

For nearly a decade after the war the development and application of nuclear energy remained largely confined to military domains. This represented a distortion in the application of scientific and technological advancements. However, the establishment of the military nuclear industrial system laid the groundwork for the subsequent transition to civilian nuclear power.

2.2.3.2 Development of first-generation nuclear power plants

2.2.3.2.1 Reactor design selection

Beginning in the mid-1950s, countries, including the Soviet Union, the United States, the United Kingdom, France, and Canada, successively established experimental nuclear power plants. Through research on multiple reactor designs, they explored the feasibility of various technical approaches. Through repeated comparisons, certain reactor designs were gradually abandoned. Although experimental in nature, the successful operation of these plants demonstrated the technical feasibility of generating electricity using nuclear power, ushering in a new era of peaceful nuclear energy utilization. The reactor designs selected for first-generation nuclear power plants were all nonproliferation reactors, primarily using natural uranium containing only about 0.7% ^{235}U as fissile fuel.

2.2.3.2.2 Soviet Union's development

Building upon modifications to military graphite-water reactors the Soviet Union constructed the world's first small-scale nuclear power plant in 1954.

2.2.3.2.3 The United States' development

The United States initiated its nuclear weapons testing program in 1952. Leveraging technology from military nuclear submarine reactors, it completed its first pressurized water reactor (PWR) nuclear power plant in 1957.

2.2.3.2.4 Development in the United Kingdom

The UK developed dual-purpose reactors for electricity generation and plutonium production based on military natural uranium graphite-gas-cooled reactor technology.

2.2.3.2.5 Evaluation

For any country with limited uranium resources, relying solely on PWRs cannot enable large-scale nuclear energy development.

2.2.3.3 Construction of second-generation nuclear power plants

2.2.3.3.1 Reactor type selection

Nuclear power plants entered practical application in the 1960s, with countries primarily focusing on developing light water reactors, heavy water reactors (HWRs), and graphite reactors. Each reactor type had its advantages and disadvantages. Relatively speaking, PWRs gained widespread attention because of their simple structure, operational safety, and lower cost. Consequently, they underwent rapid development along the lines of increasing capacity, enhancing safety, and reducing costs.

2.2.3.3.2 Research on breeder reactors commences

Research into breeder reactors began as early as the 1950s in the United States and the former Soviet Union, with subsequent emphasis also placed on this area by the United Kingdom, France, Germany, Japan, and others. Starting in the 1970s several nations gradually focused on developing breeder reactors—systems that convert nonfissile resources such as uranium and thorium into fissile artificial nuclear fuels such as plutonium and uranium.

2.2.3.4 The rise of third-generation, fourth-generation nuclear power plants, and fusion reactors

2.2.3.4.1 The promotion of third-generation reactors

Nuclear energy development centers on safety, economic viability, and sustainability. Third-generation nuclear technologies (such as AP1000, Evolutionary Power Reactor (EPR), Hualong One, and Guohua One) have been promoted globally, enhancing reactor safety and accident resilience.

2.2.3.4.2 The emphasis on fourth-generation reactors

Concurrently, demonstration projects for Generation IV nuclear technologies [e.g., sodium-cooled fast reactors, molten salt reactors, high-temperature gas-cooled reactors (HTGR)] have been implemented, advancing closed nuclear fuel cycles and industrial high-temperature applications. Countries, including the United Kingdom, France, Germany, and Japan, have researched fast breeder reactors (FBR) such as sodium-cooled fast reactors, gas-cooled fast reactors, and molten salt reactors. Among these, sodium-cooled fast reactors (commonly referred to as fast reactors) have received the most widespread attention and investment globally. France, the United States, the United Kingdom, the former Soviet Union, Germany, China, Japan, Italy, and India have pursued fast reactor development either independently or through collaborations. Fast reactor technology has matured and stands on the cusp of commercial deployment, representing a safe, reliable, and economically promising reactor type. In addition, China has completed construction of a high-temperature demonstration reactor.

2.2.3.4.3 The emergence of small reactors

Small modular reactors (SMRs) have become a research hotspot because of their high safety and flexible deployment advantages.

2.2.3.4.4 The rise of fusion reactors

Several nations are advancing fusion experiments, with projects like International Thermonuclear Experimental Reactor (ITER) entering critical construction phases.

2.2.3.4.5 Expanding the comprehensive value of nuclear energy

Nuclear energy applications continue to expand in hydrogen production, space propulsion, seawater desalination, and other fields, further enhancing its overall value.

2.2.3.5 Global nuclear energy development trends

2.2.3.5.1 Development trends in the first and second worlds

2.2.3.5.1.1 US development trends The United States is extending the operational lifespans of existing nuclear power plants while accelerating the commercialization of small reactors (NuScale's first project received approval). In September 2025 the Tennessee Valley Authority (TVA) and ENTRA1 Energy—NuScale's exclusive global strategic partner—announced a collaboration to develop a 6-GW nuclear portfolio. This includes constructing six ENTRA1 Energy plants based on NuScale small reactors at TVA sites.

2.2.3.5.1.2 French development trends France is spearheading new EPR unit construction and restarting its nuclear program. By 2025 the underconstruction Flamanville EPR unit 3 entered its final phase. France plans to build at least 6, and potentially up to 14, new-generation EPR-2 units within its existing nuclear footprint to replace aging plants and ensure future energy security.

2.2.3.5.1.3 UK development trends While advancing the Hinkley Point C nuclear project, the United Kingdom is actively pursuing small reactor development. Future plans integrate small reactor technology with data center deployment to create "energy-computing complexes," ensuring energy and computing autonomy while rebuilding its nuclear industry.

2.2.3.5.1.4 EU development trends The EU classifies nuclear power as green energy to support new projects in Central and Eastern European countries. Germany is planning its first fusion reactor.

2.2.3.5.2 Third World development trends

2.2.3.5.2.1 Asia–Africa–Latin America development trends Nuclear power plant construction is expanding into the Third World and other nations, including India, Argentina, Brazil, Egypt, Africa, and Southeast Asia. These countries are actively introducing capital and technology [14] from China, Russia, the United States, France, and others through a dual approach of "independent development" and "international cooperation" to overcome energy bottlenecks and achieve low-carbon development goals.

2.2.3.5.2.2 China development trends China entered the nuclear power sector relatively late but is now experiencing rapid expansion. Numerous nuclear power plants have commenced commercial operation, including Tianwan, Qinshan, Hongyanhe, and Daya Bay. China currently has the most reactors underconstruction globally, such as the 11 units of Hualong One and Guohua One, with technology exported to Belt and Road countries like Pakistan. It is also actively researching SMR technology. Multiple provinces, including Hunan, Hubei, and Jiangxi, are actively conducting feasibility studies for inland nuclear power plant construction.

2.2.3.5.3 Overall characteristics of nuclear energy development trends

2.2.3.5.3.1 Rapid development trend Nuclear power has rapidly emerged as one of the world's pillar energy sources. In the history of global electricity development, nuclear power has achieved the fastest growth rate. Industrialized nations with relatively scarce energy resources, particularly petroleum, have accelerated nuclear power construction. Examples include the former Soviet Union, the United States, France, Germany, Japan, and Belgium.

2.2.3.5.3.2 High-tech trend Nuclear energy technology advances toward greater intelligence, safety, and economic efficiency. Cutting-edge technological achievements are applied across all stages—from reactor design to production and management. Artificial intelligence significantly extends reactor lifespans. Laser uranium separation technology not only drastically reduces enrichment costs but also enhances uranium resource utilization. Computer control systems automatically manage an increasing number of processes, whereas new productive forces like artificial intelligence continuously advance nuclear energy development. Particularly regarding safety, some nations have developed improved and passive-safety PWRs and are even researching revolutionary reactors designed to automatically shut down under any accident scenario, preventing core meltdown. Fusion energy research is also actively progressing.

2.3 Development and application of nuclear science

2.3.1 Historical development of nuclear science

2.3.1.1 Early period (late 19th century to mid-20th century)

2.3.1.1.1 Discovery of radioactivity and elements radium and polonium, opening new frontiers in nuclear science

In 1896 Becquerel discovered radioactivity. Subsequently the Curies researched and isolated radium and polonium, ushering in a new era for nuclear science.

2.3.1.1.2 Proposal of atomic nucleus models and achievement of artificial nuclear reactions

In the early 20th century, Rutherford proposed the atomic nucleus model and achieved artificial nuclear reactions in 1919.

2.3.1.1.3 Discovery of uranium fission and clarification of its physical mechanism

In 1938 Hahn and Strassmann discovered uranium fission. Subsequently, Meitner and Frisch elucidated its physical mechanism, laying the foundation for nuclear energy development and application.

2.3.1.1.4 Construction of the world's first nuclear reactor in Chicago, marking the beginning of nuclear energy utilization

In 1942 Fermi built the world's first nuclear reactor (Chicago Pile-1) in Chicago, signifying the dawn of nuclear energy utilization.

2.3.1.1.5 Successful development of the atomic bomb and its first military application

During World War II the Manhattan Project successfully developed the atomic bomb, marking the first military application of nuclear fission technology.

2.3.1.1.6 Postwar transition to peaceful applications and commercialization

After the war, nuclear science shifted toward peaceful uses. The United Nations Atomic Energy Commission was established in 1946. In the 1950s countries such as the United States and the former Soviet Union began constructing first-generation nuclear power plants, gradually advancing nuclear energy toward commercialization.

2.3.1.1.7 Evaluation

The early development of nuclear science centered on the discovery of atomic nuclei and the theoretical foundation of nuclear fission. It began with military applications before transitioning to civilian commercialization after the war.

2.3.1.2 Mid-period (mid-20th century to late 20th century)

2.3.1.2.1 Widespread application of nuclear fission technology across sectors; second-generation nuclear power achieves commercialization

Nuclear fission technology found extensive application in nuclear power generation, military, and medical fields. Nations commenced the construction of second-generation nuclear power plants [PWR, boiling water reactor (BWR), System 80+, etc.], advancing nuclear energy commercialization.

2.3.1.2.2 Third-generation nuclear power commenced R&D and demonstration applications

Aiming to significantly enhance safety, reliability, and economics, third-generation reactor designs emerged. These designs (e.g., ABWR, AP1000, and EPR) incorporated more passive safety systems, simplified designs, enhanced accident tolerance, extended design lifetimes, and improved fuel utilization.

2.3.1.2.3 Fourth-generation nuclear power R&D commences; breakthroughs in nuclear medicine

Advanced technologies, such as FBR and HTGR, entered development. Nuclear medicine achieved breakthroughs, with radioactive isotopes applied in medical imaging and cancer treatment.

2.3.1.2.4 Two major nuclear plant accidents occur, strengthening nuclear safety regulations

The Three Mile Island (1979) and Chernobyl (1986) accidents exposed nuclear safety issues, prompting global intensification of nuclear power regulation and safety technology research.

2.3.1.2.5 Preliminary progress achieved in nuclear fusion research

Advances in nuclear fusion research emerged, with tokamaks becoming the mainstream R&D device, laying the foundation for future controlled fusion.

2.3.1.2.6 Evaluation

During this period, nuclear science entered a phase of rapid advancement. Second-generation nuclear power became a significant global energy source, particularly serving as the primary power generation method in France. Although large-scale commercial deployment of third-generation technology had not yet fully commenced, substantial progress was made in conceptual design, certification, and first-reactor construction, laying the groundwork for the next stage of nuclear power development.

2.3.1.3 Present (early 21st century to present)

2.3.1.3.1 Global promotion of Generation III nuclear power technologies

Generation III nuclear power technologies (e.g., Hualong One, Guohua One, AP1000, and EPR) are being promoted globally, enhancing safety and economic viability.

2.3.1.3.2 Initial demonstration of Generation IV nuclear power

Demonstration projects for Generation IV reactors (e.g., HTGR and sodium-cooled fast reactors) are being implemented, advancing closed nuclear fuel cycles and high-temperature industrial applications.

2.3.1.3.3 Small modular reactors and advanced nuclear fuel development gain momentum

Concurrently, research into SMRs and advanced nuclear fuels has become a focal point, enhancing flexibility and safety.

2.3.1.3.4 Nuclear fusion research accelerates

Nuclear fusion research is accelerating, with international collaborative projects like ITER advancing.

2.3.1.3.5 Evaluation

Although the Fukushima accident (2011) temporarily paused progress during the rapid development of nuclear science in the first half of the 21st century, nuclear science is now accelerating toward greater safety, efficiency, and sustainability under the dual carbon strategy and the promotion of new productive forces.

2.3.1.4 Future (mid-21st century and beyond)

2.3.1.4.1 Fourth-generation reactors are expected to enter commercial application

Fourth-generation reactors are projected to enter commercial deployment. Fast reactors combined with reprocessing technology will enable efficient nuclear fuel utilization and reduce radioactive waste.

2.3.1.4.2 Small modular reactors and microreactors may play key roles

SMRs and microreactors could become integral components of distributed energy supply systems.

2.3.1.4.3 Quantum reactors under development

With the continuous advancement of quantum computing technology, quantum reactors are also undergoing active research and development.

2.3.1.4.4 Nuclear fusion poised for demonstration power generation

Nuclear fusion is expected to achieve demonstration power generation by the mid-21st century, providing near-limitless clean energy globally.

2.3.1.4.5 Evaluation

Future nuclear science development will center on clean energy, closed fuel cycles, and nuclear fusion. The mutual advancement of nuclear science and artificial intelligence will continue optimizing global energy structures while playing vital roles in aerospace, deep-sea exploration, medicine, and AI.

2.3.2 Development of the atomic bomb

2.3.2.1 Establishment of theoretical foundations

2.3.2.1.1 Proposal of atomic models

In 1897 British physicist Joseph John Thomson discovered the electron and proposed an atomic model revealing that atoms consist of a positively charged central nucleus and negatively charged electrons.

2.3.2.1.2 Unveiling the complexity of the atomic nucleus

In 1911 British physicists Nathan Robertson and Hans Geiger discovered alpha particles and confirmed they were helium nuclei, further revealing the complexity of the atomic nucleus.

2.3.2.1.3 Discovery of nuclear fission

In 1919 British physicist Ernest Rutherford discovered nuclear fission through experiments bombarding nitrogen gas with alpha particles, proving that atomic nuclei are not indivisible.

2.3.2.1.4 Discovery of the neutron with mass identical to the proton

In 1932 British physicist James Chadwick discovered the neutron—an electrically neutral particle possessing the same mass as the proton—laying a crucial foundation for subsequent nuclear fission reactions.

2.3.2.1.5 First observation of nuclear fission

In 1934 Austrian physicists Lise Meitner and Otto Hahn observed during experiments that uranium-235 atoms, upon absorbing a single neutron, would split into two smaller nuclei while releasing vast amounts of energy and neutrons. This marked the first definitive discovery of nuclear fission.

2.3.2.2 Development and production of the atomic bomb

2.3.2.2.1 Establishment of the Manhattan Project in the United States

In 1942 the United States launched the Manhattan Project, assembling the world's foremost scientists and engineers to develop the atomic bomb. Led by General Groves with Oppenheimer as chief scientist, the project concentrated global scientific expertise.

2.3.2.2.2 Resolving key technical challenges: Nuclear materials, raw materials, and fission reactions

In 1942 Fermi led the construction of the world's first nuclear reactor, providing essential nuclear materials for atomic bomb production. Through extensive experimentation and research, scientists overcame critical technical hurdles, including the extraction and enrichment of uranium-235 and the control of fission reactions.

2.3.2.2.3 The first atomic bomb test

On July 16, 1945 the United States successfully conducted the first atomic bomb test at the Alamogordo Test Site in New Mexico. This bomb [15] had a nuclear yield equivalent to approximately 19,000 t of TNT, releasing immense energy.

2.3.2.2.4 Evaluation

The outbreak of World War II accelerated the development of the atomic bomb. Nazi Germany's demonstration of formidable military strength early in the war prompted Western nations such as the United States and Britain to intensify their atomic bomb research.

2.3.2.3 Applications and impact of the atomic bomb

2.3.2.3.1 Atomic bombings of Hiroshima and Nagasaki

On August 6 and 9, 1945 the United States dropped the atomic bombs "Little Boy" and "Fat Man" on Hiroshima and Nagasaki, Japan, respectively, causing massive casualties and property damage. The use of these bombs prompted the Japanese government to announce unconditional surrender, hastening the end of World War II.

2.3.2.3.2 Current nuclear weapons possession

Currently, multiple nations possess nuclear weapons. Beyond the United States the United Kingdom, France, Russia, China, India, Pakistan, Israel, and North Korea are also suspected of having nuclear arsenals. The existence of these weapons poses a grave threat to world peace. On June 21, 2025 the United States bombed Iranian nuclear facilities, citing Iran's nuclear activities as justification.

2.3.2.4 Evaluation

The advent of the atomic bomb marked the dawn of the nuclear age, posing immense threats to human survival and development. It spurred nations to prioritize nuclear weapons research and defense capability building, intensifying the nuclear arms race. Consequently the international community has consistently endeavored to advance nuclear disarmament and nonproliferation efforts to safeguard global security and stability.

2.3.3 Development and application of nuclear power plants

2.3.3.1 Verification and demonstration phase (1950s to early 1960s)

2.3.3.1.1 Military-to-civilian transition

2.3.3.1.1.1 Utilizing existing military nuclear technology to develop reactors for power generation During this phase, nations, including the United States, Soviet Union, United Kingdom, and France, began partially redirecting nuclear energy toward civilian use. They leveraged existing military nuclear technology to develop reactors specifically for electricity generation.

2.3.3.1.1.2 The United States constructed the world's first "artificial nuclear reactor" In 1942 a group of scientists led by Enrico Fermi in the United States built the world's first "artificial nuclear reactor," achieving a controlled, self-sustaining chain fission reaction of uranium nuclei. However, it was primarily used for scientific research and experimentation at the time and was not directly employed for power generation.

FIGURE 2.1 Obninsk graphite-moderated reactor nuclear power plant.

FIGURE 2.2 Shipinggang pressurized water reactor nuclear power plant.

2.3.3.1.2 First-generation nuclear power plants constructed

In December 1951 the experimental breeder reactor-1 (EBR-1) in the United States generated electricity from nuclear energy for the first time, marking the dawn of nuclear power history. This event not only represented a significant milestone in nuclear energy technology development but also ushered in a new chapter where nuclear power transitioned from theoretical research to practical application. The world's first commercial graphite-moderated water-cooled nuclear power plant was the Obninsk plant (now Obninsk) in the former Soviet Union, completed in 1954, as shown in Fig. 2.1. Building upon submarine reactor technology, the United States constructed the world's first commercial nuclear power plant, the Shippingport PWR plant, in 1957, as depicted in Fig. 2.2.

The completion of the Obninsk nuclear power plant (Fig. 2.1) holds profound historical significance. It not only demonstrated the feasibility of nuclear power generation but also laid the technological foundation for subsequent nuclear power development. The successful operation of this reactor marked humanity's entry into a new era of peaceful nuclear energy utilization.

The Port Peipsi PWR nuclear power plant (Fig. 2.2) paved the way for light-water reactor development, ushering in the era of peaceful nuclear energy utilization.

2.3.3.1.3 Advancing toward safer reactor designs

Following the Obninsk reactor the former Soviet Union continued developing graphite-moderated water-cooled reactor technology, which was deployed in multiple nuclear power plants. However, as nuclear technology advanced and

FIGURE 2.3 Three Mile Island nuclear power plant accident, United States.

safety standards heightened, graphite-moderated water-cooled reactors gradually revealed certain safety hazards. For example, design flaws in graphite-moderated reactors were identified as a key factor in the Chernobyl nuclear accident. Consequently, subsequent nuclear power development saw graphite reactors gradually replaced by safer, more efficient reactor designs.

2.3.3.2 Period of rapid development (mid-1960s to early 1980s)

2.3.3.2.1 Rising industrial electricity demand drives ambitious national nuclear power programs

During this period, as economies in industrialized nations grew, electricity demand surged rapidly. Countries, such as the United States, the Soviet Union, the United Kingdom, and France, all formulated large-scale nuclear power development plans.

2.3.3.2.2 The United States experienced the first peak in nuclear power plant construction

After the economic viability of American light-water reactors was demonstrated, the United States saw the first peak in nuclear power plant construction. In 1967 orders for nuclear power plants reached 25.6 GW. Following the 1973 oil crisis, efforts to reduce dependence on Middle Eastern oil triggered a second wave of nuclear plant construction, with orders totaling 66.9 GW in 1973 and 1974. Nuclear equipment manufacturing capacity reached 25–30 GW annually. The United States further established light-water reactors [16] as the dominant reactor type globally by exporting Light Water Reactor (LWR) technology and opening the uranium enrichment separation services market.

2.3.3.3 Stagnation phase (early 1980s to early 21st century)

The 1979 Three Mile Island accident in the United States and the 1986 Chernobyl accident in the former Soviet Union are illustrated in Figs. 2.3 and 2.4.

These two severe nuclear accidents significantly impacted global nuclear power development. Public concerns about nuclear safety increased, leading to

FIGURE 2.4 Chernobyl nuclear power plant accident, Soviet Union.

FIGURE 2.5 Fukushima nuclear power plant accident, Japan.

the suspension or postponement of many new nuclear power plant construction projects. To ensure nuclear safety, countries strengthened safety facilities and established stricter approval systems. The economic competitiveness of nuclear power construction declined, particularly because of uncertainties in investment risks, which hindered further nuclear power development.

2.3.3.4 Recovery and development phase (since the 21st century)

2.3.3.4.1 Delayed recovery following the Fukushima accident

With the global economic recovery and intensifying energy crises, nuclear energy regained favor as a clean energy source. However, the Fukushima nuclear accident (2011) profoundly impacted global nuclear power development, as illustrated in Fig. 2.5.

Although the Fukushima accident slowed nuclear power's pace of development, nations, learning from this painful experience under the "dual carbon" goals, have intensified efforts to advance nuclear energy recovery while simultaneously increasing focus on nuclear safety and technological improvements. Following the accident, multiple nations reassessed nuclear safety standards and advanced the upgrade of next-generation nuclear technologies to enhance resilience against extreme accidents.

FIGURE 2.6 Olkiluoto 3 EPR nuclear power plant, Finland.

2.3.3.4.2 Third-generation nuclear technology standards lead the way

Advanced light-water reactors developed by the United States, Europe, China, Japan, and other nations—representing third-generation nuclear technology—have achieved significant progress in safety, economics, and efficiency. These nuclear power plants strictly adhere to the United States. User requirements document (URD), European user requirements (EUR) document, or Chinese user requirements (CUR) document, establishing higher safety design standards. These include both active and passive safety systems, along with comprehensive engineering validation, testing, and preliminary design. Following the Fukushima accident, passive safety technologies have become a focal point, ensuring core cooling capability even during loss of off-site power and effectively reducing severe accident risks.

2.3.3.4.3 Practical development of third-generation nuclear power

Currently, multiple countries worldwide plan to expand nuclear power development, with projected sustained growth in nuclear electricity generation [17]. Post-Fukushima, Japan has gradually restarted nuclear plants while strengthening safety regulations. Europe is promoting a nuclear renaissance after establishing stricter safety standards. The United States is accelerating the commercialization of advanced nuclear technologies. China has proposed and actively implemented initiatives aligned with the Belt and Road initiative, promoting the "going global" strategy for nuclear power and establishing nuclear cooperation with multiple countries and regions [18,19]. Domestically in China, nuclear power has become a key option in energy structure adjustment, with governments at all levels increasing their focus and investment in nuclear energy. Currently, Europe has completed the Olkiluoto 3 EPR nuclear power plant in Finland, as shown in Fig. 2.6; China has constructed and operates multiple nuclear power plants utilizing third-generation nuclear technology, such as the Hualong One in Zhangzhou, the AP1000 [20–22] in Sanmen, and the EPR in Taishan, as depicted in Figs. 2.7–2.9.

With the completion of the third-generation nuclear technology "Hualong One," China's nuclear power construction capabilities have reached a globally

FIGURE 2.7 Hualong One nuclear power plant, Zhangzhou.

FIGURE 2.8 Sanmen AP1000 nuclear power plant.

FIGURE 2.9 Taishan EPR nuclear power plant.

leading level. Concurrently the "Hualong One," as China's independently developed third-generation nuclear technology route [23,24], has also been widely applied and promoted both domestically and internationally.

2.3.4 Other developments and applications in nuclear science

2.3.4.1 Energy supply

As the energy crisis intensifies, nuclear energy—a clean and efficient power source—has become a focal point for scientific research. Future breakthroughs in nuclear fusion technology hold promise for resolving humanity's energy challenges. The advancement of artificial intelligence also relies on nuclear energy for power generation.

2.3.4.2 Nuclear power plant operation optimization

AI computing power can analyze nuclear power plant operational data to optimize reactor power output and cooling system efficiency in real time. Machine learning algorithms predict equipment failures for proactive maintenance, enhancing plant safety and economic viability.

2.3.4.3 Nuclear fuel cycle management

AI computing power optimizes nuclear fuel utilization and waste management through simulation and data analysis, improving fuel cycle efficiency. Intelligent algorithms also reduce nuclear waste generation, lessen environmental impact, and advance sustainable nuclear energy development.

2.3.4.4 Nuclear-renewable energy synergy

AI computing power coordinates grid scheduling between nuclear power and renewable sources like wind and solar, ensuring stable and efficient energy supply. Combined with biomass energy for hydrogen production, it enables optimal integrated energy utilization. Through intelligent forecasting and dynamic adjustments, it maximizes clean energy utilization while reducing carbon emissions.

2.4 Development and application of nuclear technology

2.4.1 Evolution of nuclear technology

2.4.1.1 Early period (late 19th century to mid-20th century)

2.4.1.1.1 Radiation began being used for nondestructive testing

In industry, radiation was employed for nondestructive testing, enhancing equipment reliability.

2.4.1.1.2 Radiation mutagenesis breeding provided new pathways for crop improvement

In agriculture, radiation mutagenesis breeding offered novel approaches for crop enhancement.

2.4.1.1.3 Pioneering early medical imaging and the nascent development of radiotherapy

The discovery of X-rays led to early medical imaging, whereas research on radioactive elements propelled the initial formation of radiotherapy techniques.

2.4.1.1.4 Radioisotopes for analysis and tracer studies

In scientific research and environmental monitoring, radioisotopes were employed for analytical and tracer experiments.

2.4.1.1.5 Development characteristics

From the late 19th to early 20th century, nuclear technology transitioned from theoretical research to practical application. Though still nascent, this phase laid the foundation for nuclear technology's application across multiple fields, revealed its interdisciplinary potential, and paved the way for subsequent large-scale adoption.

2.4.1.2 Mid-period (mid-20th century to late 20th century)

2.4.1.2.1 Widespread adoption of radiation detection, irradiation processing, and material testing technologies

In industry, technologies, such as radiation detection, irradiation processing, and material testing, were extensively applied, enhancing manufacturing and quality control standards.

2.4.1.2.2 Gradual expansion of irradiation breeding, food irradiation preservation, and insect sterilization techniques

In agriculture, technologies, including irradiation breeding, food irradiation preservation, and insect sterilization, were progressively implemented, improving the quality and safety of agricultural products.

2.4.1.2.3 Widespread adoption of nuclear medicine equipment like PET and SPECT

Rapid medical advancements saw widespread adoption of nuclear medicine devices such as positron emission tomography (PET) and single photon emission computed tomography (SPECT), improving diagnostic accuracy and treatment efficacy.

2.4.1.2.4 Development characteristics

Nuclear fission technology matured, with nuclear power plants progressively constructed, marking the entry of nuclear energy applications into the engineering phase. During this period the standardization and large-scale application of nuclear technology were preliminarily established, with its social value becoming increasingly significant, serving as a crucial technological pillar for multiple key industries.

2.4.1.3 Present (early 21st century to present)

2.4.1.3.1 Deep industrialization of industrial irradiation technology

In industry, irradiation sterilization processes over 10 million tons of medical and food packaging annually. Industrial computed tomography (CT) and isotope tracing technologies are widely applied in nondestructive testing for high-end manufacturing, driving innovations in material modification.

2.4.1.3.2 Coordinated development of radiation breeding and molecular breeding

In agriculture, radiation breeding combined with molecular breeding enhances crop efficiency and stress resistance.

2.4.1.3.3 Enhanced application of targeted radionuclide drugs and precision radiotherapy

In medicine, targeted radionuclide drugs and precision radiotherapy enable more effective cancer treatment.

2.4.1.3.4 Nuclear detection technology is widely applied in pollution tracking, emergency response, and waste management

In environmental and public safety, nuclear detection technology is extensively used for pollution tracking, emergency response, and waste management.

2.4.1.3.5 Development characteristics

Nuclear technology increasingly aligns with societal needs, demonstrating comprehensive and systematic service capabilities. Entering the 21st century, nuclear technology integrates with cutting-edge fields such as artificial intelligence, new materials, and biomedicine, continuously expanding its application scope.

2.4.1.4 Future (mid-21st century and beyond)

2.4.1.4.1 Breakthroughs in advanced nuclear energy industrial systems

Industrial-scale application of irradiation modification for nuclear fusion materials will be achieved. Micronuclear batteries will power deep-space probes and deep-sea equipment, enabling cutting-edge manufacturing and exploration in extreme environments.

2.4.1.4.2 Nuclear technology deep integration with precision agriculture and nanotechnology

In agriculture, nuclear technology will deeply integrate with precision agriculture and nanotechnology, driving advancements in seed industry and food safety.

2.4.1.4.3 AI-assisted nuclear imaging diagnostics and personalized therapy

In medicine, AI-assisted nuclear imaging diagnostics and personalized therapy will become mainstream.

2.4.1.4.4 Development characteristics

Future nuclear technology will increasingly emphasize intelligence, environmental sustainability, and global governance coordination. Emerging productive forces, including artificial intelligence, will drive continuous nuclear technology advancement. Nuclear technology's enabling role across industries will amplify its innovation-leading effects. It will pioneer new frontiers in deep space exploration, polar development, and marine nuclear propulsion while advancing global governance mechanisms for nuclear safety, nonproliferation, and ethics—achieving deep integration of sustainable development and international cooperation.

2.4.2 Development and application of nuclear technology in the industrial sector

2.4.2.1 Purpose

2.4.2.1.1 Enhancing performance and quality

Irradiation processing can alter the physical and chemical properties of materials to improve their performance and quality. For example, gamma rays and electron irradiation can modify the molecular structure of polymers, enhancing their heat resistance, chemical corrosion resistance, and mechanical strength.

2.4.2.1.2 Preparing specialized materials

Irradiation processing is also employed to produce specialty materials such as superconductors and nanomaterials. These applications demonstrate the diverse and rapidly expanding role of nuclear technology in industrial sectors.

2.4.2.1.3 Expanding economic scale

China's nuclear technology application industry continues to grow economically, entering a phase of rapid expansion where industrial applications account for over 50% of total output value.

2.4.2.1.4 Providing industrial technical support

From enhancing material properties to ensuring product quality, nuclear technology delivers critical technical support for industrial development.

2.4.2.2 Scope

2.4.2.2.1 Overall domain

Within the industrial sector the rapid advancement of nuclear technology applications has become a driving force for continuous innovation in new technologies, materials, processes, and methodologies.

2.4.2.2.2 Specific applications

Irradiation technology modifies polymer materials to impart advantages such as low smoke emission, high-temperature resistance, and enhanced aging resistance. Nondestructive testing devices like industrial CT scanners utilize radiation to identify defects in products, equipment, and materials, thereby improving quality. Irradiation crosslinking technology, currently widely applied in polymer modification, enables linear polymers to exhibit green, efficient, and easily controllable properties under radiation exposure. Polyethylene treated with this technology demonstrates twice the temperature resistance and 10 times the wear resistance of standard Polyvinyl chloride (PVC), finding extensive use in energy, power, transportation, communications, and construction sectors. In addition, high-performance plastics obtained through radiation processing can be used to manufacture satellite components, artificial heart valves, artificial joints, artificial corneas, and more. The International Atomic Energy Agency (IAEA) notes that nuclear technology is used to determine and evaluate the properties of different materials, measure contamination levels, sterilize and disinfect components, monitor and optimize industrial processes, and modify chemical, physical, and biological characteristics to create new materials.

2.4.2.2.3 Geographic scope

China currently operates 130 cobalt-source irradiation facilities, and domestically designed and constructed cobalt-source equipment has been exported to countries, including Thailand, Vietnam, and Malaysia.

2.4.2.3 Specific applications

2.4.2.3.1 Irradiation processing

2.4.2.3.1.1 Purpose In industrial applications, equipment for sterilization and food preservation is manufactured for irradiation processing. The adoption of irradiation processing technology has significantly enhanced the market diversity of related equipment and products.

2.4.2.3.1.2 Method It utilizes gamma rays, electron beams, and X-rays to irradiate processed objects, thereby improving their quality or performance. Irradiation processing technology features strong penetration, no residual contamination, high efficiency, and energy savings, earning it recognition as a "green" processing technology.

2.4.2.3.1.3 Examples Current cobalt-60 irradiation units exceed 3×10 Ci capacity. Industrial cobalt-60 production has become an international product, with single-unit cobalt source irradiation systems now designed for 6×10 Ci capacity. Industrial irradiation processing for electron accelerators has achieved a rapid transition from low to high energy levels.

2.4.2.3.2 Radiation modification

2.4.2.3.2.1 Purpose Radiation modification is suitable to the applications of wires and cables as well as heat shrinkable materials. In addition, radiation grafting enhances substrate properties like hydrophilicity and ion exchange capacity, altering the intrinsic characteristics of polymers.

2.4.2.3.2.2 Methods Nuclear technology improves polyolefin properties through radiation crosslinking, increasing melting points, mechanical strength, and oil resistance.

2.4.2.3.3 Industrial nondestructive testing

2.4.2.3.3.1 Purpose In nondestructive testing, it enables inspection of industrial products to ensure quality and safety.

2.4.2.3.3.2 Methods Industrial CT scanning using X-rays is an application for nondestructive testing. It detects internal defects in workpieces—such as flaws in casting and welding processes—and enables real-time, noncontact thickness measurement.

2.4.2.3.3.3 Examples X-rays and gamma rays can detect defects and cracks in metallic materials without causing any damage to the material itself.

2.4.2.3.4 Irradiation sterilization

2.4.2.3.4.1 Purpose It is applicable for disinfection and sterilization of medical devices, healthcare products, daily chemical products, etc. Irradiation sterilization plays a vital role in promoting economic and social development worldwide.

2.4.2.3.4.2 Method Irradiation sterilization is an effective method that utilizes electromagnetic waves generated by ionizing radiation to kill microorganisms.

2.4.2.3.5 Radioisotope technology

2.4.2.3.5.1 Purpose Research reactors and accelerators are also used to develop new radioisotopes for applications, including nuclear medical diagnosis and therapy, nondestructive testing, industrial radiotracers, and scientific research involving radiotracers.

2.4.2.3.5.2 Method Includes: target fabrication, target irradiation, transport of irradiated targets to processing facilities, radiochemical processing or encapsulation in sealed sources, quality control, and shipment to end-users. Radioisotope production in reactors relies on neutron capture within target materials. By bombarding targets with thermal neutrons, radioisotopes are excited or produced through fission of the target material.

2.4.2.3.5.3 Example China has made progress in commercial reactor carbon-14 irradiation production targets, with market supply commencing in 2024.

2.4.3 Development and application of nuclear technology in agriculture

2.4.3.1 Purpose

2.4.3.1.1 Cultivating new varieties

Plant irradiation mutagenesis breeding utilizes various radiation sources to induce mutations in crops. Based on breeding objectives, new materials are selected from subsequent generations. These materials may be directly bred or indirectly utilized to develop new varieties. This technology can induce beneficial mutations, offering possibilities for cultivating novel varieties.

2.4.3.1.2 Ensuring food security

Countries worldwide conduct research on plant radiation mutagenesis breeding. Currently the number and cultivation area of developed mutant varieties show robust growth. This approach significantly advances the discipline of mutagenesis breeding and contributes to safeguarding food security.

2.4.3.1.3 Extending shelf life, preserving quality and flavor, enhancing safety

Agricultural irradiation processing is a vital food treatment technology. It achieves effects such as inhibiting sprouting, delaying ripening, killing insects, sterilizing, and degrading harmful substances. This extends the shelf life of agricultural products, maintains their nutritional quality and taste, and enhances food safety.

2.4.3.1.4 Promoting sustainable agricultural development

By the end of 2025 the market scale for nuclear technology applications is projected to reach trillions. These applications demonstrate the potential and significance of nuclear technology in enhancing agricultural productivity, ensuring food safety, and promoting sustainable agricultural development.

2.4.3.2 Scope

2.4.3.2.1 Overall fields

In agriculture, nuclear technology has been widely applied in plant radiation mutagenesis breeding, agricultural product and food irradiation processing, insect radiation sterilization, and other areas, becoming a crucial scientific and technological force for transforming traditional agriculture and advancing agricultural modernization.

2.4.3.2.2 Specific applications

Taking food irradiation processing as an example, radiation sterilization technology—which uses electron beams to directly or indirectly disrupt microbial ribonucleic acid (RNA), proteins, and enzymes—offers advantages such as pollution-free operation, no chemical residues, and thorough sterilization. It can eliminate microorganisms in food to prevent outbreaks of foodborne infectious diseases. In addition, it slows the metabolic rate of fruits and vegetables, extending their shelf life and playing a vital role in ensuring food safety. Plant radiation mutagenesis breeding technology, as a vital supplement to conventional breeding methods, effectively addresses issues of monoculture and material resource scarcity in agriculture and forestry.

2.4.3.2.3 Geographic scope

China collaborates with the IAEA and the Food and Agriculture Organization of the United Nations (FAO) to advance nuclear technology research and application in agriculture. Internationally the IAEA and FAO provide long-term support for global agricultural nuclear technology applications through joint programs. They promote techniques such as crop mutation breeding, pest and disease control, and soil and water resource management—particularly in developing countries—assisting member states in developing and releasing mutant crop varieties.

2.4.3.3 Specific applications

2.4.3.3.1 Radiation breeding

2.4.3.3.1.1 Purpose To select and cultivate new varieties meeting production needs.

2.4.3.3.1.2 Method Induce heritable mutations by treating plant organs with high-energy ionizing radiation such as gamma rays, X-rays, beta rays, or neutron beams. This method significantly increases mutation rates, broadens variation ranges, and accelerates mutation stabilization.

2.4.3.3.2 Radiation processing

2.4.3.3.2.1 Purpose To prevent foodborne illnesses and the spread of pests, ensure quarantine compliance for cross-border fruit and vegetable trade. This effectively addresses challenges in sterilizing frozen and refrigerated foods that conventional methods struggle to resolve, providing robust support for food safety assurance.

2.4.3.3.2.2 Method In agricultural processing, ionizing irradiation—primarily using gamma rays, X-rays, or electron beams—is employed to treat agricultural products. This technology eliminates potential harmful microorganisms.

2.4.3.3.2.3 Examples Radiation processing technology is strongly promoted by international organizations such as the FAO of the United Nations, the IAEA, and the World Health Organization (WHO). It offers advantages, including safety, efficiency, and environmental friendliness, without requiring the destruction of original packaging.

2.4.3.3.3 Irradiation storage of agricultural products

2.4.3.3.3.1 Purpose The primary purpose of agricultural product irradiation storage is to significantly reduce postharvest losses, inhibit the spread of pests and diseases, and ensure food quality and safety through the application of nuclear technology. This technology can drive the upgrading of processing and storage industries worldwide toward greater efficiency and environmental sustainability, reduce the use of chemical preservatives, and enhance the market competitiveness and international trade adaptability of agricultural products. In addition, irradiation treatment contributes to achieving food security strategies, supports sustainable development goals, and facilitates the gradual establishment of a modern, low-loss food supply chain system.

2.4.3.3.3.2 Method Irradiation technology employs controlled doses of gamma rays, X-rays, or electron beams to treat agricultural products. By disrupting the DNA structure of microorganisms, insects, and pathogens, it effectively inhibits their reproduction and activity, thereby delaying spoilage, preventing sprouting, and extending shelf life. This cold treatment process does not significantly raise product temperature, preserving the original color, texture, and nutritional value of food. Dose control and irradiation conditions must strictly adhere to international standards to ensure a safe process with no radioactive residues.

2.4.3.3.3.3 Examples For pickled peppers, irradiation primarily targets microorganisms and molds that may exist in high-moisture content and seasonings. Moderate irradiation significantly reduces contamination risks from *Escherichia coli*, yeast, etc., extending shelf life while preserving crispness and spicy flavor. This technology is widely applied to packaged ready-to-eat products such as pickled peppers and bamboo shoots. It safeguards hygiene and safety without compromising original quality, making it suitable for large-scale production and long-distance distribution.

2.4.3.3.4 Agricultural radionuclide tracing

2.4.3.3.4.1 Purpose Agricultural radionuclide tracing has expanded into fields such as agricultural ecological conservation and origin tracing for high-quality specialty agricultural products. Chemical pesticides and fertilizers are vital agricultural inputs that play a crucial role in ensuring food security. However, their improper use not only wastes resources but also generates significant environmental pollutants. Scientific application of pesticides and fertilizers to enhance utilization efficiency, reduce and control residual contamination, and provide new measures for increasing agricultural yields remains a long-term, challenging research topic attracting widespread attention from agricultural scientists worldwide. Isotope tracer technology offers advantages unattainable by conventional methods, including high sensitivity, identification capability, and traceability.

2.4.3.3.4.2 Methods Agricultural isotope tracing research primarily employs isotope tracers to investigate the migration, transformation, transport, absorption, and metabolic patterns of traced substances within plants or animals and their surrounding environments.

2.4.3.3.4.3 Examples The FAO and the IAEA have incorporated isotope tracing studies on pesticide degradation, residues, and trends in the environment into their research programs. By 2025 the Institute of Plant Protection at the Chinese Academy of Agricultural Sciences had successfully applied isotope tracing technology to investigate pesticide transformation processes in the environment.

2.4.3.3.5 Insect irradiation sterilization

2.4.3.3.5.1 Purpose Insect irradiation sterilization is a technique that uses ionizing radiation to sterilize insects. By inducing sterility mutations in pests through irradiation, their ability to reproduce is eliminated, thereby reducing or eliminating pest damage to crops. This technology offers advantages such as high efficiency, environmental friendliness, and sustainability and has been widely applied in the control of agricultural and forestry pests.

2.4.3.3.5.2 Method The technical principle of insect irradiation sterilization involves damaging insect germ cells through ionizing radiation, thereby eliminating their reproductive capacity. This technology effectively controls pest populations without damaging the environment, reducing crop damage. In addition, as irradiation sterilization [25] produces no chemical pollution or pesticide residues, it offers high safety for humans and the environment.

2.4.3.3.5.3 Examples Insect irradiation sterilization technology has been extensively applied in the control of agricultural and forestry pests. For instance, it effectively manages populations of pests such as rice stem borers, cotton bollworms, and pine caterpillars while preventing the invasion and spread of certain insects. Furthermore, this technology can be integrated with other approaches. Combining it with transgenic techniques, e.g., further enhances its pest control efficacy. For instance, Xi Zhiyong's team at Sun Yat-sen University successfully suppressed and eradicated the *Aedes albopictus* population on an island by integrating cytoplasmic incompatibility technology based on the symbiotic bacterium *Wolbachia* with radiation-induced insect sterility, demonstrating the feasibility of this approach.

2.4.3.3.6 Food safety and control

2.4.3.3.6.1 Purpose To enhance food safety and quality standards through nuclear technology applications.

2.4.3.3.6.2 Methods Detect or eliminate harmful residues and contaminants in food.

2.4.3.3.7 Plant breeding and genetics

2.4.3.3.7.1 Purpose Develop knowledge for plant breeding to create climate-resilient varieties that optimize crop yield and quality.

2.4.3.3.7.2 Methods Conduct theoretical and practical research on nuclear technology through computational and experimental approaches.

2.4.4 Development and application of nuclear technology in medicine

2.4.4.1 Purpose

2.4.4.1.1 Diagnosis and treatment

The purpose of nuclear technology in medicine is to utilize ionizing radiation from radioactive isotopes for diagnosis and treatment. Nuclear medicine—the application of radioactive isotopes in medical fields—integrates nuclear technology, electronics, computer science, chemistry, physics, and biology, forming a vital component of radiological diagnostics and therapy.

2.4.4.1.2 Conducting medical research

Nuclear medicine diagnostic and therapeutic methods also serve as vital tools for medical research. Typically, before new drugs enter clinical trials, they are labeled with radioactive isotopes to study their metabolic patterns.

2.4.4.1.3 Promoting applications and development in medicine

China has released the "Medicinal Isotope Medium-to-Long-Term Development Plan (2021–2035)," the nation's first guiding document for nuclear technology applications in healthcare. This initiative advances the establishment of a medicinal isotope supply assurance system and drives the development of novel radiopharmaceuticals and high-end diagnostic and therapeutic equipment. These advancements demonstrate the broad application and rapid progress of nuclear technology in medicine, spanning from radioactive isotope production to innovations in high-end diagnostic and therapeutic devices. Nuclear technology has made significant contributions to enhancing the precision, efficiency, and safety of disease diagnosis and treatment.

2.4.4.2 Scope

2.4.4.2.1 General field

The application of nuclear technology in medicine spans over a century. Its development in the medical field primarily focuses on two aspects: disease diagnosis and treatment. With technological advancements, nuclear technology continues to evolve toward higher imaging accuracy, stronger therapeutic targeting, and lower radiation risks.

2.4.4.2.2 Specific applications

The integration of radioactive isotopes with imaging technologies—such as PET and SPECT—enables early, precise diagnosis of tumors, cardiovascular diseases, and other conditions. In treatment, radionuclide therapy is widely used for targeted interventions in hyperthyroidism, cancer, and other diseases, enhancing therapeutic outcomes while minimizing damage to healthy tissues.

2.4.4.2.3 Geographic scope

Europe, America, Japan, and South Korea lead in high-end equipment adoption, while developing countries accelerate imports, though regional disparities remain pronounced. China's PET/CT installations rank second globally, with domestically produced equipment exceeding 40% market share.

2.4.4.3 Specific applications

2.4.4.3.1 Nuclear medicine imaging technology

2.4.4.3.1.1 Purpose Serves as a critical tool for early diagnosis of major diseases, including cardiovascular disorders, neurodegenerative diseases, and cancer.

2.4.4.3.1.2 Methods Nuclear medicine imaging techniques—including magnetic resonance imaging (MRI), CT, SPECT, and PET—noninvasively capture molecular-level information about tissues or organs.

2.4.4.3.2 Radiopharmaceutical therapy

2.4.4.3.2.1 Purpose Utilizes alpha, beta, gamma rays, and other particle beams generated by radioactive isotopes to treat pathological tissues such as tumors.

2.4.4.3.2.2 Methods Treats pathological tissues like tumors, e.g., iodine-131 therapy for thyroid diseases and thyroid cancer.

2.4.4.3.3 Boron neutron capture therapy (BNCT) for cancer treatment

2.4.4.3.3.1 Purpose Achieve industrial-scale production for treating head and neck cancer, melanoma, osteosarcoma, breast cancer, and other diseases while reducing treatment costs.

2.4.4.3.3.2 Method Treatment can be performed using compact high-current cyclotrons designed for boron neutron capture therapy.

2.4.4.3.3.3 Example In 2023 the China Institute of Atomic Energy successfully developed China's first compact high-current cyclotron, which can be widely applied in precision radiation therapy for cancer, preparation of radiopharmaceuticals, and preclinical research.

2.4.4.3.4 PET/CT imaging technology

2.4.4.3.4.1 Purpose As a high-end medical imaging examination method.

2.4.4.3.4.2 Method PET/CT integrates PET with CT technology to form an advanced medical imaging modality.

2.4.4.3.4.3 Example The Nuclear Medicine Department of Quzhou People's Hospital in Zhejiang Province successfully performed the region's first prostate cancer PET/CT imaging procedure, marking a significant advancement in precision diagnosis and treatment for prostate cancer in the area.

2.4.4.3.5 Gamma-beam stereotactic radiotherapy system

2.4.4.3.5.1 Purpose To advance the high-quality development of China's gamma knife industry and provide superior radiotherapy equipment for cancer patients.

2.4.4.3.5.2 Method Based on an innovative gamma beam stereotactic radiotherapy system.

2.4.4.3.6 Next-generation molecular imaging technology platform

2.4.4.3.6.1 Purpose To achieve breakthroughs in preclinical, clinical, and prospective research in nuclear medicine, enabling efficient and high-quality diagnosis and treatment.

2.4.4.3.6.2 Method Utilizing the next-generation molecular imaging technology platform, specifically United Imaging Healthcare's uExcel Technology and the all-core, fully digital PET/CT system uMI Panorama.

2.4.5 Development and application of nuclear technology in other fields

2.4.5.1 Biochemical field

2.4.5.1.1 Purpose

Radioisotopes serve as tracers, assisting scientists in studying chemical reaction pathways and metabolic processes within living organisms.

2.4.5.1.2 Example

Researchers commonly use carbon-14-labeled glucose to track the radioactive distribution of its metabolic products in cells or animal models. This enables the analysis of carbon flow in glycolysis and the tricarboxylic acid cycle, as well as energy production pathways.

2.4.5.2 Environmental science

2.4.5.2.1 Purpose

Nuclear technology is employed to monitor and assess environmental pollution.

2.4.5.2.2 Example

By analyzing radionuclides in the atmosphere, water bodies, and soil, scientists can trace the sources and pathways of pollutants, providing scientific basis for environmental policy formulation.

2.4.5.3 Archaeology

2.4.5.3.1 Purpose

The application of nuclear technology in archaeology offers new perspectives for studying human history.

2.4.5.3.2 Example

The use of carbon-14 isotopes in radiocarbon dating enables archaeologists to more accurately determine the chronology of ancient civilizations, thereby enhancing our understanding of the developmental trajectory of human civilization.

2.5 History and development of nuclear safety culture

2.5.1 Evolution of nuclear safety culture

2.5.1.1 Early phase: Focus on technology and equipment

In the early 20th century, nuclear safety primarily centered on the security of technology and equipment. At this time, nuclear energy applications were predominantly concentrated in the military sector, with safety measures emphasizing technical approaches. However, the subsequent Chernobyl and Three Mile Island accidents prompted deeper reflection on nuclear safety.

2.5.1.2 1980s: The emergence of humanism and risk management

By the 1980s, with the widespread adoption of nuclear energy, it became clear that nuclear safety extended beyond technical dimensions. Human behavior, decision-making, and responses were recognized as equally critical factors. Nuclear safety culture began to emphasize organizational structures, cultural values, and the role of individual behavior in shaping nuclear safety.

2.5.1.3 Early 21st century: Nuclear safety culture at organizational and individual levels

Entering the 21st century, nuclear safety culture increasingly emphasized development at both organizational and individual levels. Efforts focused on education and training, knowledge sharing, information dissemination, and reflective learning to enhance overall nuclear safety. The Fukushima accident further highlighted the importance of considering the international implications of safety culture.

2.5.2 Early history of nuclear safety culture

2.5.2.1 Interest in fission discovery and breakthroughs in nuclear theory

2.5.2.1.1 Interest in fission discovery

Scientists' engagement with nuclear energy evolved from early exploratory curiosity through gradual research and development over an extended period. Their fascination with nuclear energy originated from investigations into atomic nucleus structure and behavior. Advances in quantum mechanics and particle physics gradually unveiled the mysteries within atomic nuclei, sparking profound scientific interest in nuclear energy.

2.5.2.1.2 Breakthroughs in nuclear theory

In 1939 German scientists Hahn and Strassmann published evidence of uranium undergoing fission when bombarded by neutrons. This discovery immediately captured the attention of scientists worldwide. French scientists Joliot-Curie and others swiftly followed up, experimentally demonstrating that not only uranium but also thorium undergoes fission. They further speculated that other elements might exhibit different types of fragmentation. These studies laid the theoretical foundation for nuclear energy development.

2.5.2.2 The spirit and confidence in chain reaction research

2.5.2.2.1 Exploring methods to control chain reactions for peaceful nuclear energy utilization

Collaborating with Halban, Kowalski, and others, the Joliot-Curie team successfully revealed the generation of neutrons during fission and the potential for chain reactions. They discovered that neutrons produced by fission could trigger further fission reactions, forming a chain reaction that released immense energy. This discovery revealed the enormous potential of nuclear energy as a new energy source. Scientists began exploring how to control chain reactions to achieve peaceful nuclear energy utilization.

2.5.2.2.2 Securing multiple patents to advance nuclear development

To protect their research achievements and propel nuclear energy development, scientists applied for numerous patents. These covered various aspects of nuclear energy production, including reactor design, fuel processing, and radiation protection. Although the technological capabilities of the time were far from realizing these patented concepts, the scientists' pioneering spirit and confidence in nuclear energy's future laid the groundwork for subsequent advancements.

2.5.2.3 Evolution of public perception and attitudes

2.5.2.3.1 Lack of in-depth understanding leads to concerns and anxiety

Before nuclear technology became widespread, most people lacked a thorough understanding of it. Because of its potential for immense destructive power (as seen in nuclear weapons), the public often felt fear and resistance toward it. Furthermore, early nuclear accidents further intensified public concerns and unease about nuclear energy.

2.5.2.3.2 Weighing potential risks against benefits

2.5.2.3.2.1 Recognizing nuclear energy's role in energy conservation and emissions reduction With technological advancements and the maturation of nuclear technology, public understanding gradually deepened. People began to recognize nuclear energy's positive contributions as a clean energy source in alleviating energy crises and reducing greenhouse gas emissions.

2.5.2.3.2.2 Recognizing the balance between risks and benefits Governments, scientists, and media have intensified efforts to disseminate nuclear knowledge and promote public understanding. Through science education, public lectures, and media coverage, the public has gradually come to appreciate nuclear energy's importance and safety. In this process, people have learned to weigh its potential risks against its benefits.

2.5.2.3.3 Rational perspectives on nuclear development and application

People have begun to scrutinize nuclear technology's safety, reliability, and long-term environmental impacts, forming their own judgments and choices accordingly. Although some maintain cautious attitudes toward nuclear energy, the majority now approach its development and application with rationality.

2.5.3 Evolution and advancements in modern nuclear safety culture

2.5.3.1 Driven by technological progress

2.5.3.1.1 Origins of nuclear power technology

First-generation nuclear power plants were constructed in the 1950s, primarily serving experimental purposes. Examples include the Obninsk nuclear power plant (5000 kW capacity) built by the Soviet Union in 1954 and the Shippingport

Prototype Nuclear Power Plant (90,000 kW capacity) constructed by the United States in 1957. These plants marked the beginning of peaceful nuclear energy utilization and laid the foundation for subsequent technological applications.

2.5.3.1.2 Evolution of nuclear power technology

Second-generation nuclear power plants represented technological upgrades over the first generation. A key feature of third-generation plants was the integration of passive safety technologies, significantly enhancing safety characteristics. Multiple reactor designs emerged, including PWRs, BWRs, and HWRs, with power generation capacity substantially increasing to over 300,000 kW. Nuclear power plant construction rapidly expanded globally during this period, accelerating the commercialization of nuclear power technology.

2.5.3.1.3 Continuous technological innovation

Over time, nuclear power technology has continuously innovated, achieving breakthroughs in areas such as improving power generation efficiency, enhancing safety, and reducing construction costs. For example, the development of new reactor types such as fast neutron breeder reactors, HTGR, and SMRs has provided new directions for the future development of nuclear power technology.

2.5.3.2 Construction and safe operation of nuclear power plants

2.5.3.2.1 Global distribution of nuclear power plant construction

Nuclear power plants are constructed and operated worldwide, spanning both developed and developing countries. Countries, such as the United States, France, and Russia, lead in nuclear power, possessing extensive nuclear power plants and rich operational experience. Simultaneously, developing nations, such as China, India, and Pakistan, are actively advancing nuclear power construction to meet growing energy demands.

2.5.3.2.2 Safety and economic efficiency in nuclear power plant operation

With continuous technological progress and improved management standards the operational safety and economic efficiency of nuclear power plants have significantly enhanced. Countries have established stringent standards and regulations governing the design, construction, operation, and maintenance of nuclear power plants to ensure their safe and stable operation. For example, in September 2025, China's Ministry of Ecology and Environment is developing new regulations for nuclear power plant operator safety reports, initiating public consultation on revisions to the "Guidelines for Nuclear Safety Reports of Nuclear Power Plant Operators (Revised Draft for Comment)" issued in 2020. Simultaneously, nuclear power, as a clean and efficient energy form, holds significant importance for alleviating the global energy crisis and reducing greenhouse gas emissions.

2.5.3.3 Evolution of nuclear safety culture

2.5.3.3.1 Emergence of nuclear safety culture

The construction and operation of nuclear power plants have sparked widespread public concern regarding nuclear safety. To effectively prevent and respond to nuclear accidents, countries have intensified efforts to build nuclear safety cultures. Nuclear safety culture refers to the collective values, behaviors, regulations, and technical approaches adopted by organizations and individuals involved in nuclear activities. It encompasses norms, principles, standards, technologies, management practices, and oversight mechanisms, serving as the fundamental safeguard for nuclear safety.

2.5.3.3.2 Expansion of nuclear safety culture into broader fields

With the continuous advancement and expanding application of nuclear power technology, nuclear safety culture has gradually permeated and extended into broader domains. For instance, nuclear technology is increasingly applied in fields such as medicine, scientific research, and industry, driving progress and development in these sectors.

2.5.4 Actions and outcomes in nuclear safety culture development

2.5.4.1 Practical actions for nuclear safety culture

2.5.4.1.1 International

2.5.4.1.1.1 Policy and regulatory practices Internationally, nuclear safety culture development is advanced through multilateral treaties such as the convention on nuclear safety and standards/guidance documents issued by the IAEA, which encourage countries to strengthen nuclear safety policies and regulations.

2.5.4.1.1.2 Operational practices Internationally, regular peer reviews, experience sharing, and capacity-building projects drive continuous improvement in nuclear safety management and operational practices across nations.

2.5.4.1.2 China

2.5.4.1.2.1 Policy and regulatory practices Since 1995 China has intensified efforts in safety oversight of nuclear facilities and nuclear safety culture development. Beginning in 1997, multiple regulations mandated the establishment of procedures and systems to promote safety culture. The implementation of the Nuclear Safety Law in 2018 incorporated nuclear safety culture development into legal requirements for the first time. The 2025 implementation of China's Energy Law also explicitly recognized the vital role of nuclear energy and its safety. In September 2025, China solicited public feedback on two national ecological and environmental standards: the Draft General Guidelines for Nuclear Safety Culture Development and the Draft Implementation Guidelines for Nuclear Safety Culture Assessment. Notably the atomic energy law, promulgated on September 12, 2025, with implementation set for January 15, 2026, lays a solid foundation for future nuclear energy governance practices.

2.5.4.1.2.2 Practical implementation In advancing nuclear safety culture, China issued the Policy Statement on Nuclear Safety Culture and implemented targeted campaigns to promote and enforce safety culture requirements. Through training and awareness initiatives, all personnel are encouraged to develop legal consciousness and responsibility awareness, ensuring the effective implementation of nuclear safety culture.

2.5.4.2 Achievements in nuclear safety culture development

2.5.4.2.1 International

Regarding the achievements and outlook of nuclear safety culture development, an international nuclear safety culture system has been widely established, significantly enhancing the overall safety level of nuclear facility operations. The successful construction and operation of nuclear power plants globally because modern times have not only driven transformations in energy structures and technological innovation but also profoundly influenced the evolution of nuclear safety culture. The emergence and continuous development of nuclear safety culture provide robust safeguards for the healthy advancement of the nuclear power industry.

2.5.4.2.2 China

China has achieved initial success in building its nuclear safety culture, forming a three-tiered training system encompassing enterprises, industries, and regulatory bodies. Nuclear safety culture has become an essential and critical component of nuclear power enterprise operations.

2.5.5 Future development and outlook of nuclear safety culture

2.5.5.1 Continuous strengthening of nuclear safety culture

With ongoing technological advancements and the expanding application of nuclear energy the development of nuclear safety culture will become increasingly vital. Future efforts may further intensify publicity and education initiatives to enhance societal awareness and emphasis on nuclear safety. Simultaneously, more advanced safety management concepts and technical approaches may be introduced to reinforce the effectiveness of nuclear safety culture development.

2.5.5.2 *Innovative development of nuclear science and technology culture*

The nuclear science and technology sector will continue to embrace opportunities for innovative development in the future. With the integration of cutting-edge technologies like artificial intelligence and big data, nuclear science and technology culture will increasingly emphasize technological innovation and digital transformation. Interdisciplinary collaboration and cross-sectoral exchanges will likely intensify, driving deeper integration and synergistic innovation between nuclear science and other technological fields. As the nuclear energy industry enters a new phase of high-quality development, nuclear safety culture faces fresh developmental demands. Moving forward, it will continue leveraging artificial intelligence and new high-quality technologies to elevate nuclear safety standards, ensuring the safe and sustainable advancement of nuclear energy.

2.5.5.3 *Continuous improvement of nuclear regulatory culture*

As nuclear energy applications evolve and the international landscape shifts, nuclear regulatory culture will undergo ongoing refinement and updates. Concurrently, efforts to promote and popularize nuclear regulatory culture will intensify, enhancing societal awareness and compliance with nuclear regulations.

2.5.5.4 *Deepening development of nuclear international cooperation culture*

In the future, nuclear international cooperation culture will continue to deepen. Cooperation and exchanges with international organizations may be strengthened to advance the internationalization of nuclear regulatory culture. As global energy structures adjust and issues like climate change become increasingly prominent, cooperation among nations in the nuclear energy sector will grow closer and broader. Future efforts will continue to leverage innovative management, enhance international collaboration, and address emerging challenges to drive the sustained deepening of global nuclear safety culture.

2.5.5.5 *Trends in nuclear safety culture evolution*

The evolution of nuclear safety culture has heightened awareness of nuclear energy risks, driving the implementation of new safety standards. Integrating modern technologies—such as artificial intelligence and intelligent monitoring systems—alongside advancements in new productive forces has effectively elevated nuclear safety management levels. Concurrently, vigilance against novel high-tech risks is essential, alongside proactive promotion of global nuclear safety governance.

2.6 Nuclear technology development and direction

2.6.1 Fields and development

2.6.1.1 Fields

Since the mid-20th century, nuclear technology has undergone rapid development from fundamental research to widespread application. Globally, nuclear energy—as a clean and efficient energy form—has been extensively utilized in power generation, medicine, industry, and scientific research. For instance, China possesses a complete industrial chain and innovation chain system in nuclear technology applications, covering nearly 40 specialized fields, including medical isotopes, accelerators, radiation sources, nuclear instrumentation, irradiation applications, and distinctive nuclear medicine.

2.6.1.2 Development

According to the IAEA, by 2024, the global fleet of operational nuclear reactors exceeded 400 units, with a total installed capacity reaching hundreds of gigawatts (GW). Nuclear power generation accounted for approximately 10% of the world's electricity supply. As of June 2025 the world had 417 operational nuclear power units with a total installed capacity of 377 GW and 62 units underconstruction with a total capacity of 65 GW. By July 2025, 31 countries or regions operated nuclear power units (with 5 countries or regions—Lithuania, Taiwan, Kazakhstan, Italy, and Germany—having decommissioned all their nuclear units). The number of countries operating nuclear power units remains largely unchanged from 40 years ago (30 countries or regions had nuclear power plants in 1985). Among these 31 countries, 8 are currently constructing new units (Iran, China, South Korea, Slovakia, Pakistan, India, the United Kingdom, and Russia). By the end of 2025 the global nuclear technology application industry exceeded $850 billion in output value, driven by diverse sectors, including healthcare and nuclear power. Europe and the United States dominate in high-end equipment and technological innovation. By the end of 2024, China's nuclear technology application industry had reached nearly 900 billion yuan, maintaining an average annual growth rate above 15%. By around 2030, China's overall nuclear technology application industry is projected to double in scale. As a vital component of new-quality productive forces the nuclear technology application industry represents a quintessential strategic emerging sector. To further advance the modernization of China's nuclear technology application industrial chain the Second Nuclear Technology Application Industrial Chain Collaborative Action Conference convened in Nanning, Guangxi, on November 27, 2025. The conference unveiled China's first "Action Plan for Strengthening the Nuclear Technology Application Industrial Chain," along with the "Nuclear Technology Application Industry Yearbook 2025 Edition," the 2025 Nuclear Technology Application Industrial Chain Supply and Demand List, and the CNNC Integrated Nuclear Medical Solutions. These achievements mark China's nuclear technology application industry entering a new phase of high-quality advancement driven by systematic chain-strengthening initiatives [26].

2.6.2 Issues and challenges

2.6.2.1 Issues

Although nuclear technology applications bring benefits to humanity, they also pose significant challenges. These primarily include the risks of radioactive contamination, potential threats to the environment and human health, and the grave issue of nuclear weapons proliferation—a matter of global security that directly impacts the fundamental conditions for human survival and peace.

2.6.2.2 Challenges

The deepening application of nuclear technology also faces numerous complex challenges. For instance the long-term safe storage of nuclear waste remains a global conundrum; once nuclear accidents occur, their catastrophic consequences are often irreversible and far-reaching; the high costs associated with nuclear energy projects; and the resulting crises in public acceptance, ethics, and social trust. These collectively constitute major obstacles that must be overcome for the sustained development of this field.

2.6.3 Status and trends

2.6.3.1 Status and role

2.6.3.1.1 Multidomain energy supply and nuclear technology applications

Nuclear technology is advancing rapidly worldwide, demonstrating immense potential and application prospects across multiple domains. Significant progress has been achieved in nuclear energy, nuclear propulsion, nuclear weapons, and nuclear technology applications. It also powers artificial intelligence and is poised to maintain its position of rapid development in the future.

2.6.3.1.2 Promoting energy structure transformation and technological progress

Nuclear technology, particularly nuclear energy, serves as a "stabilizing force" in maintaining energy structure stability. Through continuous technological innovation and international cooperation, nuclear technology will play an increasingly vital role in global energy transition and technological advancement.

2.6.3.2 Development trends

2.6.3.2.1 Enhanced safety

Future nuclear technology development will prioritize safety. This includes designing safer reactors, such as third- and fourth-generation reactors with passive safety systems, to mitigate impacts from human error and natural disasters on nuclear facilities.

2.6.3.2.2 Economic optimization

With technological advancements and economies of scale the cost of nuclear power generation is expected to decrease further, enhancing its competitiveness in the energy market.

2.6.3.2.3 Environmental friendliness

As a low-carbon energy source, nuclear power will play an increasingly vital role in future energy structures. Concurrently, progress in nuclear waste treatment and storage technologies will mitigate potential adverse environmental impacts.

2.6.3.2.4 Technological diversification

Beyond traditional nuclear fission technology, research into nuclear fusion technology is actively advancing. Fusion energy, valued for its clean and sustainable characteristics, is considered the ultimate solution for future energy needs.

2.6.3.2.5 Expanded applications of nuclear technology

The application of nuclear technology in fields such as medicine, industry, and agriculture will continue to expand. Examples include using radioactive isotopes for cancer treatment, material modification, and food sterilization.

2.6.3.2.6 Policy and regulatory support

Governments worldwide will continue enacting policies and regulations to support nuclear technology development, ensuring the safe, efficient, and sustainable advancement of nuclear energy.

2.6.3.2.7 AI empowering nuclear technology

The convergence and mutual enhancement of artificial intelligence and nuclear technology represent a highly promising future direction, poised to deliver significant transformations in safety improvements, performance optimization, and cost reduction. For instance:

Real-time sensor data analysis via machine learning enables precise fault diagnosis and early warning;

Reinforcement learning allows reactors to adaptively adjust operating parameters, boosting efficiency and safety;

AI-driven optimization of nuclear fuel design improves utilization rates;

Image recognition and algorithm optimization enhance monitoring and management across fuel cycle stages; neural networks can assess nuclear facility safety and provide early risk warnings; intelligent algorithms can optimize radiation monitoring networks and protective measures; deep learning can accelerate nuclear physics experimental data processing, aiding in the discovery of new patterns; and artificial intelligence can also assist in constructing more precise theoretical models for nuclear technology.

2.6.3.2.8 Strengthened international cooperation

In the field of nuclear technology, international cooperation is increasingly vital. By sharing research findings, technologies, and experiences, countries can collectively address various future challenges in nuclear energy development.

References

[1] W.C. Röntgen, A new form of radiation, Science 3 (72) (1896) 726–729.

[2] W.C. Röntgen, On a new kind of rays, Science 3 (59) (1896) 227–231.

[3] NobelPrize.orgThe Nobel Prize in Physics 1901, Nobel Media AB, 2021 https://www.nobelprize.org/prizes/physics/1901/.

[4] Nobel Foundation, in: Nobel Lectures, Physics, 1901–1921, Elsevier Publishing Co., Amsterdam, 1967, pp. 48–72.

[5] P. Curie, Isotopy and the isotopes, J. Phys. Radium 4 (11) (1923) 381–412.

[6] Altschuler, R. Daniel, F.J. Ballesteros, in: The Women of the Moon: Tales of Science, Love, Sorrow, and Courage, 2019, pp. 182–199.

[7] NobelPrize.orgThe Nobel Prize in Physics 1903, Nobel Media AB, 2021 https://www.nobelprize.org/prizes/physics/1903/.

[8] L. Badash, Ernest Rutherford, Encyclopedia Britannica, 2024 https://www.britannica.com/biography/Ernest-Rutherford Accessed 26 August 2024.

[9] K.N. Yan, Development of the theory of microstructure of matter and modern science (Part I), Mod. Phys. Knowl. 23 (01) (2011) 50–55, doi:10.13405/j.cnki.xdwz.2011.01.012.

[10] K.N. Yan, Development of the theory of microstructure of matter and modern science (Part II), Mod. Phys. Knowl. 23 (02) (2011) 51–56, doi:10.13405/j.cnki.xdwz.2011.02.018.

[11] J.C. Pei, Y. Qiang, C.Y. Qiao, Nuclear fission—an ongoing exploration, Physics 51 (11) (2022) 770–777.

[12] M.L. Ren, B. Hu, Understanding nuclear energy, China Sci. Technol. Educ. 2020 (4) (2020) 54–55.

[13] M. Baird, Key issues in the development of nuclear energy in Asia, Nat. Environ. Law Rev. 2012 (3) (2012) 47–53.

[14] R.B. Frost, The promise of nuclear power, Int. J. Nucl. Gov. Econ. Ecol. 1 (2) (2006) 145–150.

[15] S. Davydov, Y.T. Wang, I pressed the main button for the first atomic bomb explosion—recollections of a participant in the first atomic bomb test of the former Soviet Union, Aviat. Knowl. (02) (2000) 6.

[16] Breaking the world record for continuous operation of light water reactors in the United States, Foreign Nucl. News (10) (1985) 15.

[17] J.P. Cai, W.Z. Zhang, Discussion on the technical route of pressurized water reactor nuclear power based on reflections on the development of pressurized water reactor nuclear power in the United States, in: Proceedings of the Third Radiation Protection Forum and the First Nuclear Safety Forum in the Early 21st Century, Shanghai Nuclear Engineering Research and Design Institute, China, 2004, p. 19.

[18] H.W. Song, Prospects for the development of nuclear power in China, Power Gener. Equip. (11) (1988) 23.
[19] X.P. Fang, F.G. Zhao, Application and development analysis of modular construction in nuclear power plants, in: Report on the Progress of Nuclear Science and Technology in China (Volume 4)—Proceedings of the 2015 Annual Conference of the Chinese Nuclear Society, Volume 3 (Nuclear Power Volume II), Chinese Nuclear Society; State Nuclear Power Engineering Company. China, 2015, p. 7.
[20] H.S. Wu, Y. Zhang, Westinghouse Electric Company to cooperate in building AP1000 supply chain, Foreign Nucl. News 2024 (5) (2024) 7.
[21] J.W. Zhou, H.S. Wu, Second AP1000 unit in the United States connected to the grid for the first time, Foreign Nucl. News (4) (2024) 10.
[22] Y.C. Meng, X.C. Wang, Westinghouse Electric Company signs cooperation agreement for AP1000 construction project, Foreign Nucl. News (6) (2023) 10.
[23] Y.B. Wang, The path of independent innovation in the construction of nuclear power plants in China, Power Equip. Manage. (7) (2018) 89–91.
[24] China National Nuclear Corporation, China's third-generation nuclear power technology Hualong One will be connected to the grid in 2020, Environ. Technol. 38 (02) (2020) 4.
[25] X. Zheng, D. Zhang, Y. Li, et al., Incompatible and sterile insect techniques combined eliminate mosquitoes, Nature 572 (2019) 56–61, doi:10.1038/s41586-019-1407-9.
[26] H.S. Wu, D. Dai, Review of the development of the world nuclear power industry in 2018, Foreign Nucl. News (2) (2019) 14–21.

Chapter 3

Impact of nuclear science and technology on industries

Tao Zhou[a], Yao Yao[a], Chunmei Liu[a], Shang Mao[b], Wenbin Liu[a], Xiaofang Liu[a], Haolei Zhang[a] and Shilei Dun[a]

[a] *Department of Nuclear Science and Technology, Institute of Nuclear Thermal-Hydraulic Safety and Standardization, Southeast University, Nanjing, Jiangsu, China,* [b] *School of Energy and Power Engineering, Zhengzhou University of Light Industry, Zhengzhou, Henan, China*

3.1 Impact of nuclear science and technology on industry

3.1.1 Industrial materials

3.1.1.1 Development of new nuclear fuel materials

3.1.1.1.1 High-assay low-enriched uranium (HALEU)

Centrus Energy Corp. is producing high-assay low-enriched uranium (HALEU) at a demonstration project at its American Centrifuge Plant in Piketon, Ohio, United States. In 2023 Centrus partnered with Oklo Inc. to explore the development of a comprehensive HALEU fuel cycle, including the production, deconversion, and fabrication of advanced nuclear fuel for Oklo's Aurora reactors. Concurrently, this would make Centrus a potential power offtaker for Oklo's planned microreactor in Southern Ohio. In 2024 Lightbridge Corporation initiated a Front-End Engineering and Design (FEED) study led by Centrus to add a dedicated Lightbridge Pilot Fuel Fabrication Facility (LPFFF) at the site, to determine infrastructure and licensing requirements, and to estimate costs and construction schedules. Since 2025 Lightbridge has also deepened its strategic cooperation with Oklo to evaluate the feasibility of colocating a commercial-scale fuel fabrication line within Oklo's proposed advanced fuel manufacturing facility, aiming to support fuel supply for both fast reactors and light-water reactors.

3.1.1.1.2 Enhanced low-enriched uranium (LEU+) fuel pellets

To enhance the economics and safety of nuclear power plants, Westinghouse Electric Company has developed enhanced low-enriched uranium (LEU+) fuel

Nuclear Science, Technology, and Civilization. DOI: https://doi.org/10.1016/B978-0-443-13941-3.00003-3
Copyright © 2026 Elsevier Inc. All rights are reserved, including those for text and data mining, AI training, and similar technologies.

pellets. These fuel pellets utilize Advanced Doped Pellet Technology (ADOPT). By increasing the enrichment level of ^{235}U and adding specific additives, they achieve higher safety performance and economic efficiency compared to traditional LEU fuel. The development of LEU + ADOPT fuel pellets not only improves the operational efficiency of nuclear power plants but also provides new ideas for innovation in nuclear fuel materials.

3.1.1.1.3 Integrated fast reactor metal fuel

In the field of metal fuel manufacturing, an experimental line for metal fuel manufacturing principles, led by the China Institute of Atomic Energy, completed integrated verification at the Huludao base in December 2025. This experimental line achieved principal verification of three core units—fuel slug casting, single rod fabrication, and component assembly—under simulated hot cell conditions. Among these, continuous casting technology was applied for the first time in the preparation of metal fuel slugs, potentially reducing slug seams and simplifying the process; a vertical single rod fabrication device enhanced the energy conversion efficiency of fuel elements through multidegree-of-freedom processing and sodium bonding technology; and automated vertical assembly technology optimized component production efficiency. Concurrently the team, for the first time in China, completed the preparation of a coating resistant to fuel-cladding chemical interaction on the inner surface of full-size metal fuel cladding, solving the problem of chemical interaction between metal fuel and cladding and providing key technical support for the development of integrated fast reactor metal fuel.

3.1.1.1.4 Accident-tolerant fuel (ATF) cladding materials

The International Atomic Energy Agency (IAEA)'s QUENCH-accident-tolerant fuel (ATF) joint project aims to evaluate the performance of promising ATF cladding materials under extreme conditions. Through experiments and numerical simulations, the project tests the performance of new cladding materials under extreme conditions such as high temperature and high pressure to ensure they meet the safety requirements of nuclear power plants [1]. The development of ATF cladding materials provides safer and more reliable fuel cladding options for nuclear power plants, helping to enhance their overall safety.

3.1.1.2 Application in industrial material modification

3.1.1.2.1 Radiation modification technology

Utilizing radiation sources such as neutrons and gamma rays generated by nuclear reactors, industrial materials can undergo modification treatment. Radiation modification technology can alter the microstructure of materials, thereby improving their physical and chemical properties, such as increasing strength, wear resistance, and corrosion resistance [2]. The application of radiation modification technology in the field of industrial materials promotes the research, development, and production of new materials, enhancing the quality and performance of industrial products.

3.1.1.2.2 Isotope tracer technology

Isotope tracer technology is a microanalytic method that uses radioactive isotopes as tracers to label the subject of study. This technology can track changes in materials during processing and use, providing important information for material research and development. The application of isotope tracer technology [3–6] in the field of industrial materials helps understand material performance changes, optimize processing techniques, and improve product quality and reliability.

3.1.1.3 Nuclear waste treatment and resource utilization

3.1.1.3.1 Nuclear waste vitrification technology

Nuclear waste contains a large amount of radioactive material, posing a threat to the environment and human health. Vitrification is a treatment method that immobilizes nuclear waste within a glass matrix. This technology mixes nuclear waste with glass-forming materials through high-temperature melting to form a stable vitrified waste form, effectively isolating the radioactive substances. The development and application of vitrification technology [7–9] provide an important means for the safe treatment of nuclear waste, helping to reduce its hazards to the environment and human health.

3.1.1.3.2 Recovery and reuse of elements like uranium and plutonium from nuclear waste

Nuclear waste contains significant amounts of useful elements such as uranium and plutonium, which have potential recovery value. Through advanced separation techniques and chemical processing, elements such as uranium and plutonium can be recovered from nuclear waste and reused. The recovery and reuse of elements like uranium and plutonium [10] from nuclear waste not only help reduce resource waste but can also provide a sustainable source [11] of raw materials for the nuclear energy industry.

3.1.1.4 Impact of nuclear science and technology on industrial materials in typical countries or regions

3.1.1.4.1 The United States

The United States is one of the leading countries in the global nuclear science and technology field. The impact of its nuclear technology on industrial materials is primarily reflected in the development of new nuclear fuel materials, the application of radiation modification technology, and nuclear waste treatment and resource utilization. The research, development, and application of these technologies drive innovation and development in the US industrial materials sector.

3.1.1.4.2 France

France is one of the countries with the highest proportion of nuclear power generation globally. The impact of its nuclear technology on industrial materials is mainly seen in the supply of nuclear fuel materials and nuclear waste treatment. France has a well-established nuclear fuel supply system, providing a stable and reliable fuel source for nuclear power plants worldwide. Simultaneously, France has achieved significant results in nuclear waste treatment, providing strong support for the sustainable development of the nuclear energy industry.

3.1.1.4.3 China

In recent years, China has made remarkable progress in the field of nuclear science and technology, particularly in the research and development of new types of annular fuel materials, the application of nuclear technology, and nuclear waste treatment. The development and application of these technologies have not only enhanced the competitiveness of China's nuclear energy industry but also provided strong support for innovation and development in the industrial materials sector. At the same time, China actively participates in international nuclear science and technology cooperation and exchange, making significant contributions to advancing global nuclear technology development.

3.1.1.5 Evaluation

Through efforts in the development of new nuclear fuels and materials, the application of nuclear technology in industrial material modification, and nuclear waste treatment and resource utilization, nuclear science and technology provide crucial support for the innovation and development of industrial materials. Concurrently, there are certain differences in the impact of nuclear technology on industrial materials across different countries or regions, which are closely related to factors such as their level of nuclear technology development, industrial base, and policy environment.

3.1.2 Equipment and instruments

3.1.2.1 Accelerator technology

3.1.2.1.1 Large Hadron Collider (LHC)

The Large Hadron Collider (LHC) at the European Organization for Nuclear Research (CERN) in Geneva, Switzerland, is shown in Fig. 3.1.

As can be seen from Fig. 3.1, although the LHC is primarily used for nuclear physics research, its accelerator technology provides important references for the development of industrial equipment and instruments. Accelerator technology is widely applied in fields such as medicine and industrial inspection; for example,

FIGURE 3.1 Large Hadron Collider (LHC).

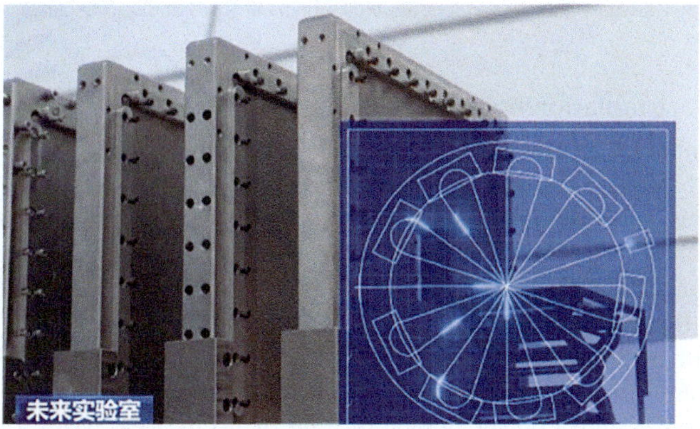

FIGURE 3.2 Petal accelerator.

medical linear accelerators are used in tumor therapy, and industrial accelerators are used in nondestructive testing, material modification, and other aspects.

3.1.2.1.2 "Petal Accelerator"

The "Petal Accelerator," designed by the Institute of Fluid Physics of the China Academy of Engineering Physics (CAEP), serves as the core device for high-energy industrial computed tomography (CT), as shown in Fig. 3.2.

As can be shown from Fig. 3.2 the "Petal Accelerator" possesses high resolution and strong acceleration capabilities, enabling precise inspection of industrial equipment and instruments. The application of this technology enhances the efficiency and accuracy of industrial inspection, providing strong support for high-end equipment manufacturing.

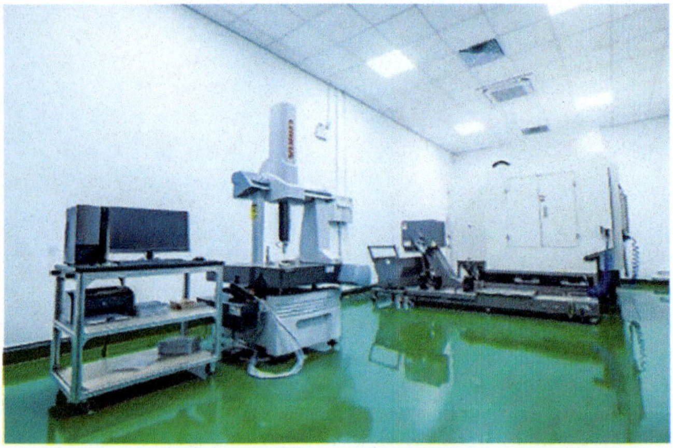

FIGURE 3.3 Advanced linear accelerator core component R&D and testing center.

3.1.2.1.3 Irradiation linear accelerator

In December 2025 the China Institute of Atomic Energy, under the China National Nuclear Corporation (CNNC), successfully and independently developed a new generation of general-purpose 10 MeV electron irradiation linear accelerator, with technical indicators reaching the international advanced level. The Advanced Linear Accelerator Core Component R&D and Testing Center is shown in Fig. 3.3.

As can be shown from Fig. 3.3, this accelerator achieves a pulsed beam current intensity of up to 400 mA, an accelerating structure capture efficiency of up to 84%, and a power utilization rate of up to 75%. With a duty cycle of only 8‰, the beam power can reach 32 kW. Leveraging core advantages such as high beam power, high capture efficiency, and high power utilization, this equipment can provide more economical, stable, and efficient solutions for large-scale production in scenarios such as food and drug irradiation, material modification, agricultural irradiation, and medical sterilization.

3.1.2.2 Nuclear imaging technology

3.1.2.2.1 Medical cyclotron

China's first medical cyclotron, developed by the team from the Institute of Fluid Physics, CAEP, has been put into operation. It is used for producing radioactive medical isotopes, enabling clear imaging of lesions under positron emission tomography (PET)–CT. The medical cyclotron not only improves the accuracy and efficiency of medical diagnosis but also provides new ideas for nondestructive testing of industrial equipment and instruments. Through similar technologies, precise imaging and inspection of industrial equipment can be achieved, enhancing product quality and safety.

3.1.2.2.2 Neutron imaging technology

The "Luoshu" neutron spectrometer neutron imaging equipment, developed by the China Mianyang Neutron Science Platform, can nondestructively detect the internal structure of materials, with detection accuracy ranging from 1 nm to 1 μm. Neutron imaging technology plays a significant role in the inspection and maintenance of industrial equipment and instruments. Through neutron imaging technology, precise measurement and analysis of the internal structure of equipment can be achieved, providing a scientific basis for equipment repair and improvement.

3.1.2.3 Nuclear radiation processing technology

3.1.2.3.1 Radiation modification

Using nuclear radiation for the modification treatment of polymer materials can alter their physical and chemical properties, such as improving strength, wear resistance, and corrosion resistance. This technology has been widely applied in the manufacturing of industrial equipment and instruments, enhancing product quality and performance.

3.1.2.3.2 Radiation sterilization

Nuclear radiation sterilization technology has advantages such as high efficiency, environmental friendliness, and no residues. It is widely used for the sterilization treatment of industrial equipment and instruments in fields such as medical treatment, food, and pharmaceuticals. The application of this technology improves product safety and hygienic quality.

3.1.2.4 Nuclear energy technology

3.1.2.4.1 Nuclear power generation

As an efficient and clean energy source, nuclear power generation provides a stable energy guarantee for the manufacturing and operation of industrial equipment and instruments. Simultaneously the development of nuclear power generation technology drives the development of related industrial chains, such as innovation and development in fields such as nuclear power instrumentation, nuclear-grade materials, and artificial intelligence.

3.1.2.4.2 Nuclear energy thermal utilization

Nuclear energy thermal utilization technology can provide high-temperature heat sources for the manufacturing and operation of industrial equipment and instruments, as well as for human life, such as in material heat treatment, chemical reactions, and residential heating. The application of this technology improves the efficiency of industrial production and human life, as well as product quality.

3.1.2.5 Impact of nuclear science and technology on industrial equipment and instruments in typical countries or regions

3.1.2.5.1 The United States

The United States holds a leading position in the field of nuclear science and technology. Its accelerator technology, neutron imaging technology, etc., play important roles in the inspection and maintenance of industrial equipment and instruments. At the same time the United States actively promotes the development and application of nuclear energy technology, providing a stable energy guarantee for the manufacturing and operation of industrial equipment and instruments.

3.1.2.5.2 France

France is one of the countries with the highest proportion of nuclear power generation globally. The impact of its nuclear technology on industrial equipment and instruments is mainly reflected in the application of nuclear power generation technology and the improvement of the nuclear energy industry chain. Through the development and application of nuclear power generation technology, France provides a clean and efficient energy guarantee for the manufacturing and operation of industrial equipment and instruments.

3.1.2.5.3 China

China has made remarkable progress in the field of nuclear science and technology, particularly in accelerator technology, neutron imaging technology, and other areas. The application of these technologies promotes the innovation and development of industrial equipment and instruments, improving product quality and performance. At the same time, China actively promotes the development and application of nuclear energy technology, providing broader market space and development opportunities for the manufacturing and operation of industrial equipment and instruments.

3.1.3 Environmental protection

3.1.3.1 Environmental protection through nuclear power generation

3.1.3.1.1 France

As one of the countries with the highest proportion of nuclear power generation globally, nuclear energy accounts for over 70% of its electricity supply. The use of nuclear power generation technology has significantly reduced greenhouse gas emissions. According to data from the IAEA, France avoids approximately 400 million tons of carbon dioxide emissions annually through the use of nuclear energy, equivalent to removing about 20 million small cars from the roads. Furthermore, nuclear power plants occupy relatively little land resources and do not produce large quantities of air pollutants during operation, which is beneficial to environmental protection.

3.1.3.1.2 China

China is actively developing nuclear power generation, with the number of nuclear power plants and the amount of electricity generated continuously increasing. As an important part of China's clean energy mix, nuclear power helps reduce the use of fossil fuels such as coal, thereby lowering greenhouse gas and air pollutant emissions. At the same time, nuclear power generation also promotes the optimization and sustainable development of China's energy structure.

3.1.3.2 Environmental protection through nuclear waste treatment

3.1.3.2.1 The United States

The United States also has extensive experience and advanced technology in nuclear waste treatment. By adopting technical means such as geological disposal and vitrification the United States effectively reduces the risk of environmental pollution from nuclear waste. Furthermore the United States actively promotes international cooperation and exchange in nuclear waste treatment technology, providing useful references for solving global nuclear waste treatment problems.

3.1.3.2.2 France

France invests substantial research and practice in nuclear waste treatment, striving to implement long-term storage and disposal of nuclear waste. Although nuclear waste treatment is a global challenge, France effectively reduces the potential environmental threat of nuclear waste by actively developing and implementing advanced nuclear waste treatment technologies.

3.1.3.3 Applications in industrial waste treatment

3.1.3.3.1 Germany

Germany has achieved remarkable results in applying nuclear radiation technology to environmental governance. German scientists have used nuclear radiation technology to treat industrial waste and contaminated soil, achieving good results. Through the application of nuclear radiation technology, Germany has effectively reduced the content of harmful substances in industrial waste and contaminated soil, improving environmental quality. At the same time the application of this technology has also promoted the development of Germany's environmental protection industry and technological innovation.

3.1.3.3.2 Japan

Japan's application of nuclear science and technology in industrial wastewater treatment is representative. After the Fukushima nuclear accident, Japan actively researched and applied nuclear technology to treat radioactive substances in industrial wastewater. By employing nuclear technology methods such as adsorption, precipitation, and ion exchange, Japan has effectively reduced the concentration of radioactive substances in industrial wastewater, decreasing environmental pollution from radioactive materials.

3.1.3.4 International cooperation in the field of environmental protection

The IAEA is an important international organization in the global nuclear science and technology field. The organization actively promotes cooperation and exchange among countries in nuclear science and technology. Through the promotion and coordination of the IAEA, cooperation among countries in the field of nuclear science and technology is continuously strengthened, jointly addressing the environmental challenges posed by nuclear technology. This helps promote the development of global nuclear science and technology and the progress of environmental protection efforts.

3.2 Impact of nuclear science and technology on agriculture

3.2.1 Crop gene optimization

3.2.1.1 The United States

3.2.1.1.1 Commercial promotion of genetically modified crops

Since 1996 the planting area of genetically modified (GM) crops in the United States has continuously increased. The main GM crops include corn, soybeans, and cotton. Based on research in nuclear technology-induced mutagenesis breeding, these crops, through genetic modification technology, have gained advantages such as excellent resistance, high yield, and suitability for mechanized planting. This has significantly improved the production efficiency of crops and promoted the development of US agriculture [12]. It has enhanced the competitiveness of US agricultural products in the international market and boosted agricultural exports [13,14]. It has also driven technological innovation in the seed industry and promoted the development of related industries.

3.2.1.1.2 Gene editing technology by Inari agriculture

Inari agriculture utilizes nuclear technology-based gene editing tools to "customize" seeds for farmers' specific crops and environments [15], aiming to improve crop yield, stress tolerance, and adaptability. Its technology has been planted on some farmland in the United States, and through two independent seed companies, a corn product has been launched on a limited basis. This provides farmers with more diverse seed choices, meeting planting needs in different regions [16]. It has promoted the commercial application of gene editing technology in the agricultural field, providing new impetus for the advancement of agricultural science and technology.

3.2.1.2 Europe

3.2.1.2.1 Genetically modified corn

Europe has approved various types of GM corn developed based on nuclear technology for use in food and feed, such as MON810, DP23211, and DP915635. These GM corn varieties, developed on the basis of nuclear technology-assisted breeding, possess insect-resistant and herbicide-tolerant traits, significantly increasing yield and reducing production costs. The promotion and use of GM corn have reduced the application of pesticides and chemical fertilizers, lowering environmental pollution. At the same time, it has improved the yield and quality of agricultural products. Furthermore, GM technology provides new ideas and methods for the development of agricultural biotechnology.

3.2.1.2.2 Genetically modified potato cultivation

Germany BASF SE has begun cultivating GM potatoes based on nuclear technology in three European countries: Germany, Sweden, and the Netherlands. These include the high-starch GM potato Amflora and the high-late-blight-resistant GM potato Fortuna. This has increased potato yield and quality, meeting the demands of the food industry for potato raw materials. It has promoted the application of genetic modification technology in plant breeding, providing new approaches for breeding disease-resistant and stress-tolerant crops.

3.2.1.3 China

3.2.1.3.1 Creation of aromatic fresh-eating corn

Fresh-eating corn refers to young corn with special flavor and quality. Based on research in nuclear technology-induced mutagenesis breeding, gene editing technology was used to alter the corn's genes, giving it an aromatic flavor suitable for fresh consumption. This technology has enriched corn varieties, met consumer demand for high-quality, high-value-added agricultural products, promoted the advancement of agricultural science and technology, and demonstrated the immense potential of gene editing technology in crop breeding.

3.2.1.3.2 Cultivation of crops such as high-yield rice and dwarf lodging-resistant high-yield corn

Through gene editing technology based on nuclear science, precise improvements are made to key agronomic traits of major food crops like rice and corn, effectively enhancing crop yield, stress resistance, and adaptability. For example, high-yield rice varieties, by improving seed setting rate and photosynthetic efficiency, have achieved continuous breakthroughs in yield per mu. Meanwhile, dwarf, lodging-resistant high-yield corn significantly enhances adaptability to mechanical harvesting and resistance to wind lodging, making it suitable for promotion and planting in various ecological regions.

3.2.1.4 Evaluation

Countries around the world have achieved remarkable results in crop gene optimization, promoting the development of agricultural science and technology and enhancing production efficiency. As a vital component of modern breeding systems, nuclear science not only underpins food security strategies but also contributes to pesticide reduction and sustainable agricultural development [17–20]. Although the potential impacts of new biotechnologies like genetic modification on human health require further study, the historical contribution and technical support role of nuclear science in the breeding process cannot be overlooked.

3.2.2 Agricultural product processing and disinfection

3.2.2.1 The United States

In the United States, nuclear science and technology are also widely applied in the processing and disinfection of agricultural products. Irradiation sterilization technology, in particular, is extensively used in the food industry for sterilizing various agricultural products and foods such as grains, fruits, vegetables, meat, seasonings, and traditional Chinese medicines. Irradiation sterilization technology not only enhances food safety but also resolves potential chemical residue issues associated with traditional sterilization methods. The application of this technology has driven technological progress and industrial upgrading in the food industry, providing consumers with safer and more reliable food choices.

3.2.2.2 China

3.2.2.2.1 Irradiation disinfection of pickled pepper chicken feet

Pickled pepper chicken feet, as a popular food, face challenges regarding shelf life and safety during production. Irradiation technology can effectively kill microorganisms in pickled pepper chicken feet, extending their shelf life while maintaining the original flavor and texture. Irradiation disinfection technology not only improves the safety of pickled pepper chicken feet but also solves the problem of flavor degradation caused by traditional thermal processing. The application of this technology promotes innovation and development in the agricultural product processing industry, providing consumers with safer and more delicious food options.

3.2.2.2.2 Irradiation preservation of potatoes

Potatoes are prone to sprouting and spoilage during storage, affecting quality and food safety. Irradiation technology [21] can effectively inhibit potato sprouting and microbial growth, extending their preservation period. Irradiation preservation technology not only solves the challenges in potato storage but also reduces storage costs and improves the economic benefits of agricultural products. The application of this technology strongly promotes sustainable agricultural development.

3.2.2.2.3 Irradiated wine for improved flavor

Irradiated wine is a modern process that uses ionizing radiation technology (such as cobalt-60 gamma rays or electron beams) to accelerate the aging of alcoholic beverages. Its core functions are First, Sterilization: High-energy gamma rays (e.g., 1.17–1.33 MeV rays emitted by cobalt-60) destroy the DNA/RNA of microorganisms, achieving sterilization at room temperature without chemical residues. Electron beam irradiation technology, which uses electron accelerators to generate high-energy electron streams, offers higher safety. Second, accelerated aging: Radiation energy excites water molecules in the wine, producing free radicals that promote esterification reactions between alcohols and acids, generating aroma compounds such as ethyl hexanoate and ethyl lactate, simulating the effect of 3–5 years of natural aging. Simultaneously, free radicals react with harmful substances like aldehydes, reducing the thrill of the wine. The effects are Increased efficiency: Traditional aging takes several years, whereas irradiation can achieve a similar taste in just days to weeks. Optimized quality: The wine becomes smoother and mellower, fusel alcohol content decreases, and aroma substances like ethyl acetate significantly increase. Enhanced safety: Strictly controlled radiation doses ensure the wine does not become radioactively contaminated, but excessive irradiation leading to peroxide residues must be avoided.

3.2.2.3 Other countries

Besides China and the United States, other countries are also actively exploring the application of nuclear technology in agricultural product processing and disinfection. For example, India's Bhabha Atomic Research Centre (BARC) has long been dedicated to research and promotion of nuclear irradiation technology in agricultural product preservation and the harmless treatment of agricultural waste. Brazil's National Nuclear Energy Commission (CNEN) supports the research, development, and implementation of various agricultural radiation technologies. Food irradiation centers have been established in multiple locations, widely used for sterilizing products such as cassava flour, spices, fruits, and meat, extending shelf life, and controlling parasites.

3.2.2.4 Evaluation

The application of nuclear science and technology in agricultural product processing and disinfection can effectively kill microorganisms and parasites, reduce the risk of contamination of agricultural products, and improve food safety. Through technical means such as irradiation, the shelf life of agricultural products can be extended, losses during storage and transportation reduced, and the economic benefits of agricultural products enhanced. The application of nuclear science and technology promotes the progress and innovation of agricultural science and technology, providing more advanced technical means and support for agricultural production. The application of nuclear science and technology in agricultural product processing and disinfection drives the upgrading and development of the agricultural product processing industry, providing consumers with safer, more delicious, and diverse food choices.

3.2.3 Intensive land use

3.2.3.1 Nuclear bomb to extinguish a fire

In 1963 a fire accidentally broke out at Well No. 11 of the Urta-Bulak gas field in Uzbekistan (then part of the Soviet Union), lasting for 2 years and wasting substantial natural gas resources. The nuclear energy department of the former Soviet Union proposed and implemented a plan to use a nuclear bomb to extinguish the fire. By drilling a directional well and detonating a low-yield nuclear bomb in the underground well the shock wave deformed the rock layers, sealing off the gas path and solving the gas leak at its source. After the nuclear bomb was detonated, the fire was completely extinguished within about 23 seconds, successfully resolving the gas leak issue.

3.2.3.2 Nuclear bomb to construct a reservoir

In 1965 water supply issues were severe for farms and industrial areas near the Chagan River bank in the Semipalatinsk region of the former Soviet Union, necessitating the construction of a reservoir. The nuclear energy department proposed using a nuclear bomb for reservoir construction. A 140-kt yield nuclear bomb was detonated at a depth of 178 m underground, creating an artificial lake. After the explosion a large crater approximately 430 m in diameter appeared on the surface, which later became the Chagan artificial lake. Because of meticulous calculation of the nuclear bomb's yield and the underground detonation, radiation levels dropped to near-natural levels shortly after the explosion, resulting in minimal impact on the environment and human health.

3.2.3.3 Evaluation

Typical applications of nuclear science and technology for intensive land use are relatively rare globally and are mostly concentrated in specific historical periods and specific countries. Although these applications demonstrate the potential of nuclear technology, they also remind people to use it cautiously and avoid the potential risks and consequences it may bring.

3.3 Impact of nuclear science and technology on the military

3.3.1 Embodiment of national defense capabilities

3.3.1.1 Impact on national defense capabilities

3.3.1.1.1 Influencing military strategy and the development of the defense technology industry

As a crucial yardstick for measuring a nation's defense capabilities, the importance of nuclear science and technology is not only reflected in enhancing comprehensive national power but also profoundly influences military strategy and the development of the defense technology industry.

3.3.1.1.2 Affecting national strategic deterrent forces and comprehensive defense strength

The research, development, and application of nuclear science and technology not only signify a nation's breakthrough in high-tech fields but also serve as the core cornerstone of its strategic deterrent forces and comprehensive defense strength.

3.3.1.2 Impact on the military strategic landscape

3.3.1.2.1 The birth of nuclear weapons

The emergence of nuclear weapons completely overturned traditional forms of warfare, driving the transformation of warfare from conventional conflicts to new forms potentially involving nuclear war.

3.3.1.2.2 The nuclear arms race during the Cold War

In their contest for global hegemony the United States and the former Soviet Union competed to develop nuclear weapons, such as the US "Minuteman" series of intercontinental ballistic missiles and the Soviet "Satan" heavy ICBMs. The development of these nuclear weapons not only enhanced their respective strategic deterrent capabilities but also prompted a realignment of global military power, forming a new strategic landscape that profoundly impacted the international political and security environment.

3.3.1.3 Promotion of innovation in the defense technology industry

3.3.1.3.1 Technological breakthroughs in key areas

Progress in fields such as nuclear energy development, nuclear materials research, and nuclear radiation protection provides indispensable technical support for the defense technology industry.

3.3.1.3.2 Development of cutting-edge weaponry and equipment

Breakthroughs in nuclear reactor technology provide a powerful power source for equipment such as nuclear submarines and aircraft carriers. For example, France's "Triomphant"-class nuclear submarines and the US "Nimitz"-class nuclear aircraft carriers have greatly enhanced naval operational capabilities and strategic deterrence.

3.3.1.3.3 Improvement of related industrial chains

Nuclear science and technology also promote the development of related industries, such as the nuclear fuel cycle industry and nuclear waste treatment technology, further injecting new vitality into the defense technology industry and driving the optimization and upgrading of the national defense technology system.

3.3.2 Strengthening of the armament system

3.3.2.1 Application of nuclear power technology in nuclear-powered facilities

3.3.2.1.1 Nuclear-powered submarines

The US "Virginia"-class submarines and Russia's "Borei"-class submarines, with their nearly unlimited endurance and extremely high stealth capabilities, have become the backbone forces for missions such as distant patrols, strategic deterrence, and precision strikes. It is expected that the UK's "Dreadnought"-class nuclear submarines, entering service sequentially from the 2030s, will ensure the continuation of the UK's nuclear deterrent capability into the 2060s.

3.3.2.1.2 Nuclear-powered aircraft carriers

The US "Ford"-class carriers, with their powerful aircraft launch and recovery capabilities and comprehensive combat effectiveness, serve as strategic pivots and mobile fortresses in modern naval warfare. The United Kingdom once focused on demonstrating the application of nuclear power technology in aircraft carrier platforms but later shifted its efforts to the development of conventionally powered carriers based on its national defense needs. France possesses the "Charles de Gaulle" nuclear-powered aircraft carrier, which is currently the only operational nuclear-powered carrier outside the United States, providing core support for the French Navy's distant operations. The former Soviet Union began constructing the nuclear-powered carrier "Ulyanovsk," but it was later dismantled. The former Soviet aircraft carrier "Admiral Kuznetsov" was laid down at the Nikolayev South Shipyard in the USSR on April 1, 1982; launched on December 4, 1985; commissioned on January 21, 1991; and in 2025, Russia stated that the "Admiral Kuznetsov" is no longer worth repairing. However, future aircraft carriers built by China may adopt nuclear power.

3.3.2.2 Nuclear missiles driving changes in military operational methods

3.3.2.2.1 Enhanced performance, enabling long-range and high-precision strike capabilities

Nuclear technology improves the performance of weaponry and equipment. The US "Minuteman" III ICBM possesses a range of thousands of kilometers and high-precision strike capabilities. In September 2025 the US Navy successfully test-fired four unarmed "Trident II" D5 ballistic missiles from an "Ohio"-class ballistic missile submarine in waters near Florida. Russia is deploying the second-generation, dual-capable (conventional/nuclear) "Oreshnik" intermediate-range missile in the Russia–Ukraine war, which has a range of thousands of kilometers. China's "Jinglei-1" air-launched long-range missile, "Julang-3" submarine-launched intercontinental missile, "Dongfeng-61" land-based intercontinental missile, and the new "Dongfeng-31" land-based intercontinental missile demonstrate China's "triad" strategic nuclear forces (land, sea, and air-based). The "Dongfeng-5C" liquid-fueled intercontinental strategic nuclear missile is a significant component of China's strategic counterattack system.

3.3.2.2.2 Changes in operational methods, enhancing combat effectiveness and strategic flexibility

The ability to rapidly destroy enemy strategic targets enhances the military's combat effectiveness and strategic flexibility. Test launches serve as both deterrence and practical drills. The United States test-fired a nuclear-powered missile in May 2025; Russia test-fired the new "Burevestnik" nuclear-powered cruise missile in October 2025.

3.3.2.3 Promoting development in other military technology fields

3.3.2.3.1 Radar technology

Breakthroughs in phased array radar technology enable radar systems to track and process multiple targets simultaneously, providing critical information support for air and naval warfare.

3.3.2.3.2 Quantum communication technology

Provides ultrahigh-speed, ultrasecure, and ultralong-distance transmission means for military communications, improving communication efficiency and confidentiality.

3.3.2.3.3 Global satellite navigation systems

Provide precise positioning and navigation services for military operations, supporting precision strikes and rapid maneuvering in modern warfare.

3.3.2.4 Impact on military training and combat simulation

3.3.2.4.1 Military training and combat simulation

Utilizing nuclear science and technology for tactical drills and combat simulations helps the military simulate real battlefield environments and combat scenarios, improving the troops' ability to handle complex situations.

3.3.2.4.2 Virtual reality and war gaming

Virtual reality technology is used in tactical drills to enhance soldier skills and team coordination; war gaming provides commanders with accurate battlefield situational information, supporting the formulation of scientific operational plans and decisions.

3.3.2.5 Evaluation

The application of nuclear science and technology achievements, such as nuclear power technology and nuclear missiles, has not only greatly enhanced the operational effectiveness of naval and strategic strike forces but also promoted the overall upgrade of related military technologies. Concurrently the integrated application of various advanced technologies, such as radar, quantum communication, satellite navigation, and combat simulation, provides solid technical support for modern military training and command decision-making, marking a fundamental transformation in military operational methods.

3.3.3 Deterrence of nuclear forces

3.3.3.1 Definition of nuclear deterrence

Nuclear deterrence [22], as a military strategy theory, refers to the use of nuclear forces and the threat of nuclear war by nuclear powers to handle international affairs and pursue global strategies in the context of the nuclear age.

3.3.3.2 Conditions for implementing nuclear deterrence

1. The existence of nuclear forces
2. The effectiveness of nuclear weapons
3. The resolve and will to use nuclear weapons

3.3.3.3 Nuclear deterrence as a means of strategic defense

3.3.3.3.1 Safeguarding national security

Nuclear deterrence can effectively prevent potential aggressors from hostile acts, as any use of force challenging national sovereignty and territorial integrity faces immense risks, thereby safeguarding national security and stability.

3.3.3.3.2 Promoting international peace and stability

Under the balance of nuclear deterrence, major powers can maintain strategic restraint, avoiding direct conflicts and the outbreak of war. This balance provides a stable security environment for the international community, conducive to cooperation and development among nations.

3.3.4 Control of nuclear weapons

3.3.4.1 Purpose of controlling the use of nuclear weapons

Controlling the use of nuclear weapons is a major issue of international concern, with the core purpose being to prevent the catastrophic consequences of nuclear war and maintain global peace and security[23,24]. To avert nuclear catastrophe the international community has promoted a series of landmark international treaties through diplomatic efforts.

3.3.4.2 Key international treaties

3.3.4.2.1 Treaty on the Non-Proliferation of Nuclear Weapons (NPT)

3.3.4.2.1.1 Objectives Treaty on the Non-Proliferation of Nuclear Weapons (NPT) has effectively curbed the horizontal proliferation of nuclear weapons and provided the international community with a stable nuclear security framework. For example, the Iran nuclear deal was reached within the NPT framework, serving as a model for addressing nuclear proliferation crises.

3.3.4.2.1.2 Content In effect, since 1970, the NPT is the cornerstone of the global nuclear nonproliferation regime. It stipulates that nuclear-weapon states shall not transfer nuclear weapons or related technology to nonnuclear-weapon states; nonnuclear-weapon states pledge not to seek to acquire nuclear weapons, while also emphasizing the right and obligation to peacefully use nuclear energy.

3.3.4.2.2 Comprehensive Nuclear-Test-Ban Treaty (CTBT)

3.3.4.2.2.1 Objectives To comprehensively ban all nuclear explosion tests (including for military and civilian purposes), thereby curbing the development of new nuclear weapons, preventing the escalation of the nuclear arms race, and promoting the nuclear disarmament process by establishing a global monitoring network (using technical means such as seismic and infrasound monitoring), ultimately reducing the long-term threats of nuclear proliferation risks and radioactive contamination to ecology and humanity.

3.3.4.2.2.2 Content Requires states parties to undertake to prohibit any nuclear weapon test explosion or any other nuclear explosion (including for peaceful purposes) in any environment and to establish an international verification system composed of seismic, hydroacoustic, infrasound, and radionuclide monitoring stations, supplemented by on-site inspection mechanisms, to transparently oversee compliance. Although the treaty has not yet fully entered into force, its signing represents a significant step forward for the international community in the fields of nuclear disarmament and nonproliferation.

3.3.4.3 Challenges facing nuclear weapons control

3.3.4.3.1 Geopolitics and covert means

Some countries, because of geopolitical considerations or strategic security needs, may circumvent the restrictions of international treaties. Their nuclear intentions and programs have long troubled the international community.

3.3.4.3.2 Development of new types of nuclear weapons

With the advancement of technology, the development of new low-yield tactical nuclear weapons or directed-energy weapons lowers the threshold for nuclear weapon use, further complicating nuclear proliferation and control efforts.

3.3.4.4 Strategies to address challenges in nuclear weapons control

3.3.4.4.1 Strengthening international cooperation and nuclear disarmament

Promoting the nuclear disarmament process to reduce the number and threat of nuclear weapons. For example, the now-defunct United States–Russia New START treaty was a positive step by both sides to reduce their nuclear arsenals.

3.3.4.4.2 Enhancing international oversight of nuclear materials and technology

Establishing stricter export control mechanisms and strengthening the monitoring of nuclear materials. For example, the IAEA's safeguards system ensures that nuclear weapons are not illegally manufactured or used.

3.3.4.4.3 Raising public awareness and nuclear security education

Initiatives such as the "nuclear disarmament education" promoted by institutions like the United Nations Educational, Scientific, and Cultural Organization (UNESCO) can enhance public understanding of the dangers of nuclear weapons and cultivate values of peace and nonviolence.

3.3.4.4.4 Optimizing monitoring with modern technological means

Utilizing technologies such as artificial intelligence and big data analysis can help improve nuclear nonproliferation monitoring and verification capabilities, enabling faster and more accurate identification of potential proliferation risks and providing effective early warning and response mechanisms.

3.4 Impact of nuclear science and technology on the development of medical technology

3.4.1 Advancements in nuclear medicine imaging technology

3.4.1.1 Definition and characteristics

Nuclear medicine imaging diagnosis is a method that uses nuclear technology to diagnose and treat diseases. Specifically, nuclear medicine imaging refers to a technique that uses nuclear technology methods to obtain images of the interior of the human body, offering advantages such as high resolution, high sensitivity, noninvasiveness, and no radiation (from the imaging procedure itself, although it involves radioactive tracers).

3.4.1.2 Methods and types

3.4.1.2.1 Types

Commonly used nuclear technologies in nuclear medicine imaging diagnosis include magnetic resonance imaging (MRI), CT, single photon emission computed tomography (SPECT), and PET.

3.4.1.2.2 Principles

PET and SPECT technologies use biological agents labeled with radioactive nuclides for imaging, providing information on tissue function and metabolism. MRI and CT technologies, on the other hand, use magnetic fields and X-rays to acquire images.

3.4.1.2.3 Methods

When performing nuclear medicine imaging diagnosis, doctors select the appropriate nuclear technology for examination based on the patient's condition and needs. By analyzing the images, doctors can understand information such as the location, size, shape, and quantity of the lesion and its relationship with surrounding tissues, thereby making an accurate diagnosis.

3.4.1.3 Applications and effects

These technologies can not only diagnose, stage, and monitor the efficacy of tumors but also assess neurological and cardiovascular diseases. At the same time, they can characteristically trace specific functions of organs to diagnose and evaluate diseases. Nuclear medicine imaging technology has evolved from SPECT and PET to multimodal imaging technologies like SPECT/CT, PET/CT, and PET/MR, which integrate with imaging techniques such as CT and MR, improving the accuracy and reliability of disease diagnosis.

3.4.1.4 Development and directions

The development and application of technologies such as MRI, CT, SPECT, and PET continuously drive the progress of medical imaging, providing powerful tools for disease diagnosis and treatment. With continuous technological advancement, their application scope and precision are also constantly expanding and improving. With the rapid development of artificial intelligence technology, nuclear medicine imaging diagnosis is also moving toward automation and intelligence. For example, using deep learning technology, lesions in images can be automatically identified and analyzed, improving the accuracy and efficiency of imaging diagnosis.

3.4.2 Development of nuclide therapy

3.4.2.1 Definition and characteristics

Radionuclide therapy is a method that uses radioactive nuclides and their labeled drugs for disease diagnosis and treatment. At the same time, radionuclide therapy is targeted, as diseased tissues can highly and specifically concentrate the radioactive drug, resulting in good therapeutic effects and low toxic side effects.

3.4.2.2 Methods and types

It involves labeling a carrier with a radioactive nuclide, which is then introduced into the corresponding organ of the human body, causing the radioactive nuclide to concentrate in that organ, thereby achieving a therapeutic effect through irradiation and internal radiotherapy. In nuclide therapy, besides routine treatments for diseases like thyroid cancer, bone metastases, and liver cancer, the continuous emergence of new radiopharmaceuticals targeting different tumor markers (such as PSMA, FAPI, CXCR4, DOTATATE, $\alpha v\beta 6$) labeled with imaging and therapeutic nuclides (e.g., 18F, 68Ga, 99mTc, 177Lu, 225Ac) is making the integration of nuclide diagnosis and therapy for different tumor targets a hot topic and focus of research and application, accelerating the development of nuclear oncology.

3.4.2.3 Applications and effects

Radionuclide therapy is mainly applicable to certain malignant tumors of the thyroid gland, for example, thyroid cancer, hyperthyroidism, and tumor treatments for some other endocrine glands. Through radionuclide therapy, diseased cells can be inhibited or killed to achieve therapeutic goals. This treatment method usually does not require surgery, causes less patient discomfort, and has relatively reasonable costs, making it suitable for patients who cannot undergo surgery or have financial difficulties.

3.4.2.4 Development and directions

Future nuclide therapy will develop toward higher precision targeted therapy through the development of new radiopharmaceuticals (such as alpha emitters and multitarget probes), combined with multimodal imaging technologies, to promote the widespread application of integrated personalized diagnosis and treatment models. At the same time, exploring the combined application of nuclide therapy with emerging therapies such as immunotherapy and gene therapy will become an important direction for breaking through the bottleneck in treating refractory tumors.

3.4.3 New drug development and disease evolution research

3.4.3.1 Definition and characteristics

Nuclear medicine technology also plays a vital role in new drug development, disease evolution research, and the development of new diagnostic and therapeutic techniques. Its core characteristic is the use of radioactive tracers to achieve noninvasive, dynamic, and quantitative monitoring of physiological and biochemical processes within living organisms. This allows researchers to track drug distribution, metabolic pathways, and the mechanisms of disease occurrence and development in real-time at the molecular level.

3.4.3.2 Methods and types

The development of radiopharmaceuticals requires multiple stages, including lead compound screening, pharmacodynamics studies, pharmacokinetics studies, and safety evaluations. This process involves the use of various technologies and equipment, such as radionuclide production, labeled compound synthesis, animal experiments for pharmacodynamic evaluation, and clinical trials.

3.4.3.3 Applications and effects

Radiopharmaceuticals are drugs containing radioactive nuclides and can be used for tumor diagnosis and treatment. For example, fluorodeoxyglucose (FDG) is a commonly used radiopharmaceutical for PET/CT scans, providing high-resolution, high-sensitivity images. At the same time, some radiopharmaceuticals, such as I-MIBG, can also be used for tumor therapy.

3.4.3.4 Development and directions

In the future, this field will develop toward the creation of novel molecular probes with higher sensitivity and specificity, deeply integrating artificial intelligence technology to accelerate the process of new drug screening and evaluation. Simultaneously, research will focus on achieving rapid translation from basic research to clinical application. Through the combination of multimodal imaging and precise quantitative analysis, breakthroughs in personalized medicine and dynamic disease monitoring technologies will be promoted.

3.4.4 Boron neutron capture therapy

3.4.4.1 Definition and characteristics

Boron neutron capture therapy (BNCT) is a tumor-targeted radiotherapy method based on the principle of nuclear reactions. Its core mechanism involves selectively accumulating drugs containing nonradioactive boron-10 within tumor cells; then, external irradiation with low-energy thermal neutrons triggers a reaction between boron-10 and the neutrons, producing high linear energy transfer (LET) particles (alpha particles and lithium-7 nuclei), thereby achieving precise killing of tumor cells. This technology is characterized by high selectivity, high local lethality, and relatively minor damage to normal tissues.

3.4.4.2 Methods and types

The BNCT treatment process mainly consists of two major steps: first, introducing the boron carrier drug into the body via injection, allowing it to accumulate in the tumor region; subsequently, using a reactor or accelerator to generate thermal neutrons for irradiating the target area within the body. Based on the neutron source, BNCT can be categorized into research reactor-based BNCT and accelerator-based BNCT (AB-BNCT). The latter has become a research hotspot in recent years because of its compact equipment and suitability for hospital environments.

3.4.4.3 Applications and effects

BNCT is primarily used to treat tumors insensitive to conventional treatments, such as malignant gliomas, recurrent head and neck cancers, and melanomas. Clinical studies show that BNCT has significant advantages in prolonging patient survival, alleviating symptoms, and improving quality of life, especially for recurrent lesions or those difficult to resect surgically.

3.4.4.4 Development and directions

With advancements in drug design and neutron source technology, efforts are focused on improving the targeting and selectivity of boron compounds to more precisely localize cancer cells and reduce impact on healthy tissues. Combined with multimodal treatment approaches, BNCT is gradually moving toward widespread clinical application. Currently, some research institutions in China have begun patient clinical trials, marking a significant leap for BNCT in China and its entry into a new phase.

3.5 Impact of nuclear science and technology on high-tech advancement

3.5.1 Quantum nuclear reactors and their impact on future development

3.5.1.1 The concept of a quantum nuclear reactor

Traditional nuclear reactors are based on the laws of classical physics, maintaining system stability by controlling the nuclear chain reaction. Reactor design relies on classical models of neutron transport and heat transfer. While effective, this approach faces challenges related to safety, waste treatment, and limited resources. Against this backdrop, scientists have proposed the concept of the quantum nuclear reactor. In traditional nuclear reactors, nuclear fission or fusion processes are achieved through macroscopic physical phenomena. The idea of a quantum nuclear reactor, however, might involve utilizing quantum phenomena such as quantum entanglement and superposition to optimize these nuclear reaction processes. Research in this field is continuously evolving, with scientists exploring how quantum technologies can be applied to various aspects of nuclear energy.

3.5.1.2 The goal of a quantum nuclear reactor

A quantum nuclear reactor is a novel type of nuclear reactor design based on the principles and technologies of quantum mechanics, aiming to utilize quantum effects in nuclear fusion and fission processes to enhance the efficiency, safety, and reliability of energy generation. Unlike traditional nuclear reactors, quantum nuclear reactors fully leverage the coherence and superposition properties of quantum states, along with the latest developments in quantum computing and quantum information processing, to provide new ideas and technological pathways. The quantum nuclear reactor is a relatively cutting-edge and challenging concept, combining quantum mechanics with nuclear energy technology to explore the use of quantum techniques for improving the efficiency, safety, and controllability of nuclear energy. Although research in this area is still in the developmental stage, its potential applications could significantly alter humanity's understanding and utilization of nuclear energy.

3.5.1.3 Innovative features of quantum nuclear reactors

3.5.1.3.1 Features and purpose

The innovation of the quantum nuclear reactor lies in its attempt to apply principles of quantum mechanics, such as the quantum tunneling effect, quantum coherence, and quantum decoherence, to enhance the efficiency and safety of nuclear reactions. The core concept of a quantum nuclear reactor is to utilize quantum effects like quantum tunneling, quantum coherence, and quantum decoherence.

3.5.1.3.2 Quantum tunneling effect

Traditional nuclear reactions (especially fusion reactions) require particles to overcome a certain energy barrier to proceed. The quantum tunneling effect allows particles to pass through these energy barriers, potentially making it possible to achieve fusion reactions at lower temperatures. This not only improves efficiency but also reduces reliance on extreme conditions.

3.5.1.3.3 Quantum coherence effect

The coherence property of quantum systems allows for the interference and superposition of particle states, which could create new possibilities within nuclear reactions. By manipulating quantum states, reaction conditions could be optimized, enhancing energy output.

3.5.1.3.4 Quantum computing and simulation

Using quantum computing technology the processes of nuclear reactions can be simulated and predicted with greater precision, which will positively impact the optimization of reactor design and operation.

3.5.1.4 Applications of quantum nuclear reactors

3.5.1.4.1 Clean energy production

The high efficiency and relative safety of quantum nuclear reactors make them an ideal method for clean energy production. If successful, nuclear fusion can be achieved, which would provide a nearly inexhaustible source of clean energy, reducing dependence on fossil fuels and helping to address the challenges posed by climate change. The author's team of this book has been granted a patent for a heat pipe quantum microreactor.

3.5.1.4.2 Nuclear medicine applications

Specific radioactive isotopes generated by quantum nuclear reactors could be used for medical imaging and therapy. For the monitoring and treatment of diseases like cancer, a quantum nuclear reactor could deliver quantitative amounts of radioactive isotopes as needed, offering greater precision and flexibility.

3.5.1.4.3 Fundamental scientific research

Quantum nuclear reactors could serve as important tools for research in particle physics and nuclear physics. By analyzing quantum behavior within the reactor, scientists could gain a deeper understanding of nuclear forces, elementary particles, and the formation processes of the universe.

3.5.1.5 *Future prospects of quantum nuclear reactors*

3.5.1.5.1 Fundamental research

Deepen the understanding of how quantum effects influence nuclear reactions through experimental and theoretical research, aiming to establish a foundational theoretical framework for quantum nuclear reactions.

3.5.1.5.2 Technological realization

Develop new materials and technological means to apply quantum control and monitoring techniques to practical reactors. Compared to traditional reactors, quantum reactors may require more complex control systems and materials.

3.5.1.5.3 International cooperation

Because of the global impact of nuclear energy, international cooperation and exchange will propel research progress in quantum nuclear reactors. Sharing data and technologies can accelerate the development of this new field.

3.5.1.5.4 Development challenges

Quantum nuclear reactors are still in the theoretical exploration stage, and their practical application faces numerous technical and engineering challenges. However, research in this field holds great potential and could lead to significant transformations in the nuclear energy industry.

3.5.2 Fusion reactors and their impact on the development of basic sciences

3.5.2.1 *The concept of a fusion reactor*

A fusion reactor is a device that generates energy through nuclear fusion reactions. Its core principle involves fusing light atomic nuclei (such as the hydrogen isotopes deuterium and tritium) into a heavier nucleus (such as helium), releasing a substantial amount of energy. This process is the fundamental mechanism powering stars like the Sun. The design and construction of fusion reactors aim to mimic these natural reactions for energy production on Earth. The basic equation is

$$D+T \rightarrow \mathrm{He}+n+17.6\ \mathrm{MeV} \tag{3.1}$$

where D stands for deuterium, and T stands for tritium. The reaction produces helium and a fast neutron, releasing approximately 17.6 million electron volts of energy. This type of reaction possesses an extremely high energy density, generating energy more efficiently and cleanly than fossil fuels or nuclear fission reactions.

3.5.2.2 Pathways to fusion

3.5.2.2.1 Gravitational confinement

The Sun employs the gravitational confinement pathway. Because of its enormous mass the sun generates immense gravity, enabling gravitational confinement fusion reactions under extreme temperatures and pressures. Humanity is currently unable to use gravitational confinement to achieve fusion reactions.

3.5.2.2.2 Inertial confinement

Inertial confinement fusion (ICF) uses energy from drivers (such as lasers or particle beams) to compress a deuterium–tritium fuel pellet, creating a high-temperature, high-pressure plasma. During the inertial confinement time of the high-density fuel plasma, fusion ignition and burn are achieved. The National Ignition Facility (NIF) in the United States and China's Shenguang laser facilities both follow the inertial confinement pathway.

3.5.2.2.3 Magnetic confinement

Magnetic confinement uses specially shaped magnetic fields to confine a high-temperature plasma of a certain density, composed of atomic nuclei such as deuterium and tritium and free electrons, within a limited space. This allows it to reach the fusion condition—the Lawson criterion—thereby releasing energy. Typical research pathways using magnetic confinement include the tokamak, tandem mirror, reversed-field pinch (RFP), Z-pulsed power facility, and stellarator.

3.5.2.3 Impact of fusion reactors on basic sciences

3.5.2.3.1 Particle physics and nuclear physics

Research into fusion reactors involves fundamental theories of particle physics, including the strong interaction, weak interaction, and quantum chromodynamics (QCD). Fusion reactions require understanding the behavior of nuclear forces under extreme conditions, providing an important experimental platform and theoretical foundation for studying high-energy physics.

3.5.2.3.2 Plasma physics science

The core of a fusion reactor is the generation and control of plasma. Plasma physics is a crucial branch of modern physics, covering topics such as particle behavior and wave phenomena. Fusion research has driven a deeper understanding of plasma behavior and led to the development of new theories and models, helping to explain plasma phenomena in celestial bodies like the sun.

3.5.2.3.3 Materials science

Fusion reactors operating under extreme conditions (high temperature, high radiation, etc.) impose stringent requirements on the materials used for their construction. Researchers need to develop materials capable of withstanding high-energy neutron irradiation and high temperatures, prompting research and development into new alloys, ceramics, and composites, thereby advancing materials science.

3.5.2.3.4 Energy science and sustainable development

Fusion reactors are seen as a key potential solution for future clean energy. Compared to fossil fuels and nuclear fission energy, fusion energy produces no long-lived radioactive waste, has an abundant fuel supply, and poses essentially no risk of severe nuclear accidents. Therefore its research not only promotes the development of basic sciences but is also of great significance for the research and development of sustainable energy solutions.

3.5.2.3.5 Computational science and engineering

The design and operation of fusion reactors rely on complex numerical simulations and computational engineering techniques. From simulating plasma behavior to analyzing material properties, fusion research drives progress in high-performance computing, algorithm optimization, and engineering design, promoting the multidisciplinary intersection of computational science.

3.5.2.3.6 Interdisciplinary fields

Multiple disciplines, such as polymer science, laser physics, and fluid mechanics, are closely related to fusion technology. The development of fusion reactors fosters interdisciplinary connections and spawns new research areas, such as the study of laser–plasma interactions, thereby expanding the boundaries of science.

3.5.2.4 Future prospects

Research into fusion reactors still faces technical and engineering challenges, including how to better control plasma, improve the efficiency of fusion reactions, and reduce uncertainties in modeling and materials. International collaborative projects like International Thermonuclear Experimental Reactor (ITER) demonstrate the scientific community's confidence in and efforts toward achieving controlled nuclear fusion, marking the promise of fusion reactors as a future energy source. Fusion reactors are not only an important pathway to achieving clean and sustainable energy, but their development also profoundly impacts various fields of basic science, driving progress in physics, materials science, computational science, and many other disciplines. As the technology matures, fusion energy is expected to provide a continuous source of power for humanity's energy needs in the future, contributing to the development of human society.

3.5.3 Impact of other nuclear-related scientific frontiers

3.5.3.1 Promotion of nuclear physics and particle physics

Research into nuclear energy has driven in-depth exploration of elementary particles. Through particle accelerators (such as the LHC), scientists can recreate high-energy nuclear reactions, detect elementary particles like quarks and gluons, and deepen the understanding of the fundamental constituents of matter. At the same time the study of nuclear forces in nuclear reactions influences the basic theories of nuclear physics, enabling scientists to better understand nuclear structure, including the stability of atomic nuclei and reaction mechanisms. This research provides a theoretical foundation for nuclear energy development.

3.5.3.2 Integration of nuclear energy with renewable energy

Research into nuclear energy promotes the exploration of new types of power systems, such as the integration of nuclear power with renewable energy sources like wind and solar power. This diversified energy solution helps address the challenges of global warming and energy shortages. With the continuous advancement of nuclear and renewable energy technologies, more innovative solutions are expected to emerge, such as hydrogen-based energy storage technologies and smart interconnected grids. These will contribute to more efficient energy management systems. The successful experiences of different countries in integrating nuclear and renewable energy can provide references for others, promoting international cooperation and technology transfer to jointly tackle climate change and energy scarcity.

3.5.3.3 New quality productivity and AI-driven nuclear energy innovation

The deep integration of artificial intelligence and nuclear technology is fostering new quality productivity. AI optimizes the design of fourth-generation reactors through multiphysics coupling algorithms. Combined with digital twins and sensor networks, it enables fault prediction and autonomous robotic operation and maintenance. It also uses generative adversarial networks to accelerate the development of nuclear-grade materials. Simultaneously, deep reinforcement learning in plasma control for fusion devices pushes the limits of magnetic confinement stability, driving the iteration of nuclear energy systems toward greater safety, efficiency, and lower carbon emissions. According to IAEA statistics, 73% of nuclear facilities globally have deployed AI-assisted decision-making modules, with the goal of reducing nuclear power operating costs by 40% by 2030.

3.6 Impact of nuclear science and technology on health

3.6.1 Disease diagnosis and radiation health

3.6.1.1 Development purpose and assurance

The application and advancement of nuclear science and technology in the medical field is a continuously evolving field. With the continuous development and improvement of nuclear science and technology, its application in the medical field will become more widespread and in-depth, potentially contributing more to the cause of human health. However, at the same time, nuclear science and technology also require strict safety management and supervision to ensure that their application processes do not cause harm to people or the environment. Therefore, while applying nuclear science and technology, it is necessary to strictly abide by relevant laws, regulations, and standards, strengthen safety management and technical supervision, and ensure human health and safety.

3.6.1.2 Application of nuclear science and technology in medical diagnosis

The initial application of nuclear science and technology in the medical field was in medical diagnosis, primarily utilizing nuclear medicine imaging technologies, including SPECT, PET, MRI, and CT. These have been explained in Section 3.4.

3.6.1.3 Application of nuclear science and technology in radiation health

3.6.1.3.1 Definition and scope

The application of nuclear science and technology in the medical field is not limited to medical diagnosis and treatment but also involves radiation health. Radiation health refers to the measures and methods used to protect people and the environment from the hazards of radioactive substances. The application of nuclear science and technology in radiation health mainly includes radiation detection, nuclear safety supervision, etc.

3.6.1.3.2 Application in nuclear safety supervision

Radiation detection refers to the testing and monitoring of radiation levels in people, the environment, food, etc. Nuclear science and technology can monitor radiation levels through equipment such as nuclear radiation detectors, nuclear radiometers, and nuclear radiation alarms, promptly detecting radiation accidents and taking measures.

3.6.1.3.3 Application in radiotherapy room protection

Nuclear science and technology can also be applied to the protective design and protective measures of hospital radiotherapy rooms. Nuclear safety supervision refers to the supervision and management of the safety of nuclear facilities, ensuring their operational safety and environmental safety.

3.6.1.3.4 Application in nuclear facility assurance

Nuclear science and technology can be applied to the safety design, safety assessment, and safety monitoring of nuclear facilities, providing guarantees for the safety of nuclear facilities.

3.6.2 Assisted treatment of diseases

3.6.2.1 International medical applications

3.6.2.1.1 Proton therapy in the United States

Proton therapy technology in the United States, especially at the MD Anderson Cancer Center, is one of the world's leading medical applications of nuclear technology. Proton therapy uses positively charged proton beams to provide patients with high-precision treatment, minimizing damage to surrounding healthy tissues. It excels in treating childhood cancers, head and neck cancers, brain tumors, etc., becoming an important supplementary method for cancer treatment, particularly suitable for cases requiring precise control of radiation dose.

3.6.2.1.2 Heavy ion therapy in Japan

Japan's heavy ion therapy has a significant impact globally. The National Institute of Radiological Sciences (NIRS) in Chiba, Japan, has been in operation since 1994, using carbon ion beams to treat cancer. Compared to traditional radiotherapy, heavy ion therapy offers stronger penetrating power and precision, enabling more effective treatment of deep-seated or complexly located tumors. Its therapeutic effect is particularly remarkable for cases with special needs, such as ocular cancers and brain tumors.

3.6.2.2 Medical applications in China

3.6.2.2.1 Development of the Wuwei Heavy Ion Center

The first carbon ion therapy system with independent intellectual property rights in China officially entered clinical treatment at the Wuwei Heavy Ion Center. Researchers began deep-seated tumor clinical trials in March 2009. Results show that heavy ion beams have significant advantages in improving long-term survival rates and quality of life for deep-seated tumors, especially refractory tumors that recur or are unresponsive to conventional treatment.

3.6.2.2.2 Development of "Kylin Knife" technology

2024 marks the "first year of full marketization" for CSMC (China Science and Technology Medical Corporation). The "Kylin Knife" (Qilin Knife) precision radiotherapy series products have been used in clinical treatment in multiple hospitals. "Kylin Knife" technology is an important development in the field of precision radiotherapy for tumors in China, currently featuring technological breakthroughs and independent intellectual property rights. It includes five series of products: Intensity Modulated Radiation Therapy Planning System software, medical linear accelerator, X-ray image guidance system, dose verification system software, and radiotherapy information and image management software.

3.6.2.2.3 Development of boron neutron capture therapy (BNCT)

In September 2025 Xiamen Humanity Hospital in China, in cooperation with Neuboron Medtech, completed the enrollment for the Phase I registration clinical trial of BNCT, and Phase II clinical registration trial research has been fully initiated. Since the commencement of clinical trials, Xiamen Humanity Hospital has conducted over 70 BNCT clinical research treatments cumulatively.

3.6.3 Nuclear radiation dose and health

3.6.3.1 Definition and classification of interventional radiology doses

3.6.3.1.1 Definition

According to the 2000 report of the United Nations Scientific Committee on the Effects of Atomic Radiation (UNSCEAR), doses from interventional radiology procedures have been specifically described[25]. Interventional radiology operations can potentially cause both deterministic injuries and stochastic radiation damage. In such cases, doses may be large enough to cause skin and lens injuries to patients. In interventional radiology procedures, the highest dose to the patient is at the skin entry point of the radiation (Peak Skin Dose, or PSD).

3.6.3.1.2 Classification

A 2000 report by the International Commission on Radiological Protection (ICRP) has classified interventional procedures based on the maximum cumulative absorbed dose to the patient's skin: High-dose procedures are defined as those causing doses of several hundred milligray (mGy); medium-dose refers to those causing doses of several tens of mGy, and low-dose procedures are those causing doses below 10 mGy [26]. Therefore special attention should be paid to the radiation injuries induced in patients by high radiation doses.

3.6.3.2 Dose data standards for coronary interventional procedures

3.6.3.2.1 Situation

E Vano et al. in Spain [27] studied the skin radiation doses and injuries in 14 patients who underwent 4–14 coronary angiography procedures and 5–10 percutaneous transluminal coronary angioplasty (PTCA) procedures over a period of 2–10 years.

3.6.3.2.2 Values

The estimated mean dose-area product (DAP) per procedure was 46 Gy • cm^2 for coronary angiography and 82 Gy • cm^2 for PTCA. The mean maximum skin dose per procedure was 217 mGy for diagnostic studies and 391 mGy for PTCA procedures.

3.6.3.3 Impact of radiation hazard effects on hematology and immune function

3.6.3.3.1 Characteristics

Regarding the hematological changes caused by radiation hazard effects [28], the abnormal rates of various biological indicators in the interventional radiology group were significantly higher than those in the general radiology group and the control group.

3.6.3.3.2 Manifestations

Main manifestations include neurasthenic syndrome and skin changes. There is an increased detection rate of abnormal total white blood cell count, increased chromosome aberration rate, and increased micronucleus cell rate; total white blood cell count and platelet count are below normal; lymphocyte ratio is above normal; cellular immunity T lymphocyte subsets are significantly lower than the control group; red blood cell C3b receptor activity is inhibited, and immune complex levels are increased, indicating that immune function is affected.

3.6.4 Nuclear war, nuclear radiation, and mental health

3.6.4.1 Nuclear war and mental health

3.6.4.1.1 The catastrophic threat of nuclear weapons and urgent response

Nuclear weapons are the most destructive weapons ever created. In the event of a full-scale nuclear conflict between major powers such as the United States and Russia, a 2022 study predicts global deaths could exceed 5 billion, with irreversible damage to social structures and the ecological environment. Therefore the most urgent task now is to ease major power confrontations, restart peace talks, and buy time and space to prevent catastrophic consequences.

3.6.4.1.2 Nuclear tensions and proliferation dilemmas under geopolitical games

"Nuclear tension" stems from the escalating geopolitical competition and sense of insecurity among major powers. As the risk of strategic confrontation rises, global strategic stability is rapidly eroded, falling into a vicious cycle of "seeking to be safer—becoming less safe—desiring absolute security even more." For example, the US bombing of Iranian nuclear facilities on June 21, 2025, further intensified the development of tensions. At the same time, the "nuclear double standard" policy long pursued by the United States and other Western countries has left many loopholes in the global nuclear nonproliferation regime, not only lowering the threshold for using nuclear weapons but also creating conditions for proliferation by emerging nuclear states. In November 2025 both the United States and Russian presidents attempted to restart nuclear testing, also contributing to international nuclear tensions.

3.6.4.1.3 Prospects for global strategic stability and disarmament cooperation

Maintaining global strategic stability is the fundamental prerequisite for preventing nuclear war. The joint statement by the leaders of the five nuclear-weapon states (the United States, Russia, China, the United Kingdom, France) in 2022 emphasized that "a nuclear war cannot be won and must never be fought," calling on all parties to gradually reduce the size of nuclear forces through dialogue and cooperation, provided their own security is not undermined. Only through such joint efforts can nuclear weapons be truly used solely for defense and deterrence, creating a secure environment conducive to disarmament and ultimately moving toward a world free of nuclear weapons.

3.6.4.1.4 Impact of nuclear war on public psychology

Populations living under the long-term shadow of the nuclear war threat also face serious mental health challenges. Research indicates that anxiety, helplessness, and pretraumatic stress disorder (PTSD) triggered by nuclear risks are spreading among populations in different countries, particularly affecting adolescents and vulnerable groups profoundly. Reducing nuclear risk is not only a matter of national security but also pertains to the psychological safety of all humanity.

3.6.4.2 Nuclear radiation and mental health

3.6.4.2.1 Nuclear radiation terrorist events

These refer to terrorist events where, for evil political, economic, religious, or ethnic purposes, terrorists threaten to use or actually use devices capable of releasing radioactive material (including crude nuclear explosive devices and radiological dispersal devices), or threaten to attack or actually attack nuclear facilities (including major radiation source irradiation facilities), causing the release of radioactive material, leading to psychological and social impact on a significant number of people or a certain number of casualties, thereby disrupting state functions, public life, social stability, and economic development.

3.6.4.2.2 Impact of nuclear radiation on public psychology

3.6.4.2.2.1 Psychological stress reactions Nuclear accidents or radiation events can trigger intense psychological stress reactions in individuals. For example, after the Chernobyl accident, although the expected external and internal exposure doses for the public were relatively low, the psychosocial impact was extremely significant, leading to widespread symptoms of stress, anxiety, mental disorders, depression, etc.

3.6.4.2.2.2 Mental health issues The psychosocial impact of nuclear and radiological emergencies includes anxiety, panic, long-term psychological stress, etc. These issues not only affect individuals' physical and mental health but can also lead to disruption of social life and production order. The actual impact can be far greater than the direct harm caused by the radiation itself.

3.6.4.2.2.3 Impact on specific populations After a nuclear accident, specific groups such as women, children, and cleanup workers are particularly severely affected in terms of mental health. Women and cleanup workers often experience depressive symptoms and PTSD because of long-term psychological pressure. Mothers/pregnant women in affected areas may also experience family conflicts because of cognitive differences related to radiation risk, potentially even affecting suicide rates.

3.7 Impact of nuclear science and technology on daily life

3.7.1 Educational choices

3.7.1.1 Cultivation of professional talent

The demand for professional talent in the nuclear science and technology field promotes the establishment and development of related majors in universities. For example, focusing on key areas such as nuclear power generation, nuclear propulsion, medical isotopes, radiopharmaceuticals, and nuclear medicine, universities are supported in setting up relevant undergraduate majors. This provides students with more specialized choices, particularly in fields like nuclear engineering and nuclear technology, nuclear material and chemical engineering, and nuclear medicine.

3.7.1.2 Educational curriculum system

The curriculum system for majors like nuclear engineering and nuclear technology includes general education courses, professional education courses, practical education courses, interdisciplinary modules, and personalized modules aimed at cultivating students' comprehensive quality and professional skills.

3.7.1.3 Cultivation of practical and innovative abilities

Students majoring in nuclear engineering and nuclear technology need to possess innovative abilities and the basic training capability to complete various experiments. Approximately 20% of the class hours in professional education courses are allocated to practical teaching segments, including experimental courses, various internships, and practice activities.

3.7.1.4 Foreign language and international perspective

Students in nuclear engineering and nuclear technology need to master at least one foreign language for application; be able to communicate and exchange ideas in a cross-cultural context; and possess an international perspective, enabling them to read foreign language literature in the field of nuclear engineering and technology.

3.7.1.5 Employment directions and prospects

Graduates of nuclear engineering and nuclear technology have broad employment prospects. They can work in areas such as nuclear power plants, nuclear fuel cycle and storage, nuclear safety regulation and management, and radiation environmental protection. They can also engage in nuclear technology R&D, nuclear detection, nuclear medicine imaging, etc. Furthermore, graduates can work in government agencies, university research institutions, scientific research institutes, state-owned enterprises and institutions, multinational corporations, etc., undertaking related technical work, R&D work, management work, or education and training.

3.7.1.6 Science popularization and education

The popularization of nuclear science and technology knowledge is also emphasized. Through building science popularization bases, holding open days, lectures, publicity campaigns, site visits, and other activities, outreach efforts are increased. This helps enhance public awareness and understanding of nuclear science and technology and may also influence students' interest in and choice of nuclear-related majors.

3.7.1.7 Policy support

National policies support the development of nuclear technology application industries, including enhancing industrial infrastructure services, improving talent introduction and cultivation mechanisms, and increasing industrial promotion efforts. These policies provide students in nuclear-related majors with more learning and development opportunities.

3.7.1.8 Evaluation

The impact of nuclear science and technology on educational choices is multifaceted, ranging from major offerings and curriculum systems to practical ability cultivation, employment prospects, and science education, all providing students with more diverse educational options and career development paths.

3.7.2 Career choices

3.7.2.1 Professional talent cultivation and employment prospects

The demand for professional talent in the nuclear science and technology field drives the establishment and development of relevant university majors. Nuclear engineering and nuclear technology are high-tech fields intersecting multiple disciplines, including two main branches: nuclear power engineering and nuclear energy utilization and nuclear technology and applications. This major boasts a relatively high employment rate. In addition, related majors like nuclear chemical engineering are now also gaining attention and development.

3.7.2.2 Professional skills and career development

The overall employment situation for graduates directly matching their nuclear major is characterized by a decent average outcome but a limited ceiling for top-tier earnings. The main employment direction for graduates is state-owned enterprises, such as nuclear power design, research, and operation units, nuclear fuel and waste processing plants, etc. These large State-owned enterprises (SOEs) and research/design institutes offer stable salaries and benefits, which may not be extremely high but are still a viable option, especially in an era where employment within the system is once again becoming a good choice. With the world refocusing on nuclear energy and technology the nuclear major is gaining popularity.

3.7.2.3 Interdisciplinary and cross-cutting fields

The nuclear engineering and technology major has high foundational requirements and broad interdisciplinary applications, leading to relatively diverse career paths. For example, those skilled in electronics might go to hardware departments of companies like Huawei. Although compensation might not match that of electrical engineering and computer science majors, it is generally not below the average for engineering disciplines. The current development of artificial intelligence further brings cross-cutting opportunities.

3.7.2.4 International perspective and foreign language ability

Students in nuclear engineering and technology need to master at least one foreign language for application, effectively listening, speaking, reading, and writing, enabling communication and exchange in cross-cultural contexts. This provides students with opportunities to work and develop on international platforms.

3.7.2.5 Cultivation of practical and innovative abilities

Students in nuclear engineering and technology need to possess basic experimental abilities, scientific research innovation abilities, and engineering practice abilities. The practical teaching system includes four levels: public cognitive experiments, professional basic experiments, professional application experiments, and technological innovation experiments. This helps students integrate theoretical knowledge with practice, improving their ability to solve practical problems.

3.7.2.6 Project management and lifelong learning

Students in nuclear engineering and technology need to master engineering management principles and economic decision-making methods, applying them in multidisciplinary environments. At the same time, students need to have an awareness of autonomous learning and lifelong learning, possessing the ability to continuously learn and develop.

3.7.2.7 Professional ethics and teamwork

Students in nuclear engineering and technology need to possess good humanities and social science literacy, a strong sense of social responsibility, and qualities such as rigorous scholarship, pragmatism, truth-seeking, unity, and collaboration, aiming for comprehensive development morally, intellectually, physically, aesthetically, and labor-wise.

3.7.2.8 Evaluation

The impact of nuclear science and technology on career choices is comprehensive, from the cultivation of professional skills to career development paths, from broadening international perspectives to enhancing practical innovation abilities, all providing students with diversified career options and development opportunities. Particularly in the context of the "dual carbon" strategy and the development of new quality productive forces, including artificial intelligence, students in nuclear majors face numerous important development opportunities.

References

[1] D. Peter, S. Juri, G. Mirco, et al., Analysis of iron-chromium-aluminum samples exposed to accident conditions followed by quench in the QUENCH-19 experiment, J. Nucl. Mater. 580 (2023) 154433.

[2] R. A. Osawa, B. T. Barrocas, O. C. Monteiro, et al., Photocatalytic degradation of amitriptyline, trazodone and venlafaxine using modified cobalt-titanate nanowires under UV–Vis radiation: Transformation products and in silico toxicity, Chemical Engineering Journal 373 (2019) 1338–1347.

[3] N.-N. Grazielle, S.C.-R. Caleb, V. Marina, R. Felicidad, Microwave radiation induces modifications in the protein fractions of tef flours and modulates their derived techno-functional properties, Int. J. Biol. Macromol. 253 (P4) (2023) 126908.

[4] C. Liguo, Z. Zhengchao, C. Xiaojiao, et al., Research progress on the application of plutonium isotope tracing technology in tree rings, Sci. Geogr. Sin. 44 (5) (2024) 911–916, doi:10.13249/j.cnki.sgs.20220737.

[5] N. Ben, Z. Kai, Z. Juntao, et al., Application progress of isotope tracer technique in the study of direct coal hydrogenation mechanism, J. Fuel Chem. Technol. 50 (12) (2022) 1535–1546, doi:10.19906/j.cnki.jfct. 2022050.

[6] J.L. Liao, M. Tian, Y. Du, Q. Ji, L. Zhu, X. Zhang, Y. Hua, X. Li, Application of deuterium oxide (D_2O) isotope tracing technique for encapsulated QFN failure analysis, Microelectron. Reliab. 135 (2) (2022) 114607.

[7] X. Kai, International research progress on nuclear waste glass solidification, Mater. Rev. 35 (7) (2016) 481–488.

[8] W. Yu, X. Qingli, M. Baojian, et al., Research progress on high-level radioactive waste glass solidification technology, China Build. Mater. Sci. Technol. 29 (4) (2020) 26–28.

[9] M. Ciecińska, P. Stoch, A. Stoch, et al., Thermal properties of $60P_2O_5$–$20Fe_2O_3$–$20Al_2O_3$ glass for salt waste immobilization, J Therm Anal Calorim 121 (2015) 1225–1232.

[10] L. Wang, Nuclear transmutation helps solve the challenge of nuclear waste disposal, China Energy News 011 (2024), doi:10.28693/n.cnki.nshca.2024.000962.

[11] Y. Yang, Issues and solutions in international law regarding nuclear waste disposal, Heilongjiang Hum. Resour. Soc. Secur. 2021 (21) (2021) 25–29.

[12] F. Guohua, Z. Xinyu, J. Ruijie, et al., Research progress on quality evaluation and preservation technology of fresh corn, Food Saf. Guide (11) (2023) 169–171, doi:10.16043/j.cnki.cfs.2023.11.012.

[13] Y. Zhou, L. Zhang, Analysis of the impact of genetically modified crop cultivation on agricultural development, Spec. Econ. Zone Econ. 2024 (7) (2024) 38–41.

[14] C. Aifeng, Application of genetic modification technology in crop improvement, Hebei Agric. Mach. (14) (2024) 99–101, doi:10.15989/j.cnki.hbnjzzs.2024.14.032.

[15] Y. Meng, Z. Zou, F. Meng, Current status and prospects of the genetically modified crop industry, Agric. Dev. Finance 2024 (6) (2024) 44–46.

[16] Inc. Inari Agriculture, Inari Agriculture, Inc. files SEC Form D, notice of exempt offering of securities: (Aug. 9, 2018), Agric. Week (2018) 559.

[17] L. Wen, Commercial cultivation of genetically modified bananas receives “green light”, China Science Daily 2024 (001) (2024), doi:10.28514/n.cnki.nkxsb. 2024.000408.

[18] J.R. Roberts, V. Naimy, Strategic adoption of genetically modified crops in Lebanon: a comprehensive cost–benefit analysis and implementation framework, Sustainability 16 (19) (2024) 8350.

[19] K.A. Naveen, M. Sontakke, A review on regulatory aspects, challenges and public perception in acceptance of genetically modified foods, Food Sci. Biotechnol. 33 (4) (2024) 791–804.

[20] M. Awais, Q.A. Rao, S. Sadaqat, et al., The regulatory landscape of genetically engineered crops in Pakistan, Food Energy Secur 13 (1) (2024) 1–11.

[21] Y. Jiangtao, B. Junqing, L. Kui, et al., Effects of X-ray irradiation on the storage and preservation of coloured potatoes, Sci. Technol. Food Ind. 45 (15) (2024) 322–331, doi:10.13386/j.issn1002-0306.2023080324.

[22] W. Zhengda, Mechanism of nuclear deterrence: strength foundation, signal transmission, and psychological game, Int. Forum 24 (1) (2022) 98–126, doi:10.13549/j.cnki.cn11-3959/d.2022.01.006.

[23] W. Yaqiang, The Use and Reflection of Nuclear Weapons in World War II, Northwest Normal University, Gansu, 2013, doi:10.7666/d.D359661.

[24] D. Ying, L. Bin, W. Riqiang, Taboo and arms control, World Econ. Politics 2010 (8) (2010) 48–62.

[25] United NationsSources and Effects of Ionizing Radiation: Vol. 1—Sources, 2000 Report to the General Assembly with Scientific Annexes, United Nations Publications, New York, 2000.

[26] J. Jiangtao, Avoidance of radiation injuries from medical interventional procedures, Ann. ICRP 30 (2) (2000) 7–67.

[27] E. Vano, J. Goicolea, C. Galvan, et al., Skin radiation injuries in patients following repeated coronary angioplasty procedures, Br. J. Radiol. 74 (26) (2001) 1023–1031.

[28] Y. Xinfang, S. Ximei, C. Wenhua, et al., Analysis of radiation dose and effects among interventional radiology workers in ten hospitals, Chin. J. Radiol. Health 9 (3) (2000) 177–178.

Chapter 4

Impact of nuclear science and technology on regions

Yao Yao, Tao Zhou, Dongli Huang, Peng Liu, Chunhui Xue, Jianyu Tang, Shilei Dun and Zongyan Bai

Department of Nuclear Science and Technology, Institute of Nuclear Thermal-hydraulic Safety and Standardization, Southeast University, Nanjing, Jiangsu, China

4.1 Impact of nuclear science and technology on Europe

4.1.1 General development of European nuclear science and technology

4.1.1.1 Development status

From the late 19th century to the early 20th century, it was an exciting period of nuclear scientific discovery, as European scientists discovered revolutionary new phenomena in nuclear physics such as radioactivity, nuclear structure, nuclear fission, and nuclear fusion [1]. From World War II to the period of the mid-1940s to early 1990s, nuclear technology was first used to develop nuclear weapons. Shortly after the end of World War II, European countries promoted the peaceful use of nuclear energy, and nuclear technology was widely applied in many fields such as commercial or research nuclear power plants. After the Chernobyl nuclear accident the Three Mile Island nuclear accident, and especially the Fukushima nuclear accident, some countries have opposed nuclear energy and even abandoned nuclear energy. Germany has decided to completely phase out nuclear power by 2022; Belgium has decided to exit the nuclear power program by 2025; and Italy, Spain, and Switzerland have decided not to build new nuclear power plants. On the other hand, after the conflict between Russia and Ukraine, to get rid of dependence on Russian energy, Finland's Generation III nuclear power technology, the European pressurized reactor (EPR), was connected to the grid for power generation in March 2022 and was officially put into production in April 2023. Some European countries, such as France and the Netherlands, also plan to build new nuclear power plants in the next few years. In 2025 France plans to build up to 14 EPR-2 units and by 2030 aims to increase nuclear power generation to 360–400 billion kWh per year. The Netherlands plans to construct

Nuclear Science, Technology, and Civilization. DOI: https://doi.org/10.1016/B978-0-443-13941-3.00004-5
Copyright © 2026 Elsevier Inc. All rights are reserved, including those for text and data mining, AI training, and similar technologies.

four new nuclear power stations, with the goal of having them operational by 2035.

4.1.1.2 *Development trends*

Before 2030 the Generation III nuclear power technology will remain the main force of nuclear power development in Europe. Safety remains the prerequisite for the development of nuclear power, and the overall development direction is to maximize the effectiveness of nuclear energy utilization. Around 2030 mature Generation IV reactors or advanced small reactors will be introduced to the market and then gradually expand their scale to meet the needs of multipurpose utilization of nuclear energy. Fusion power is one of the most desirable future energy sources in Europe. The superconducting Tokamak International Thermonuclear Experimental Reactor (ITER), which is supposed to produce controllable and large-scale nuclear fusion reactions, is currently under construction in Europe. Germany is also intending to develop nuclear power, specifically announcing plans to build the world's first fusion power plant, and has released Europe's first conceptual design for a commercial nuclear fusion power station.

4.1.2 Specific development of nuclear science and technology in Europe

4.1.2.1 *Scientific research field*

European scientists have mainly relied on the European Organization for Nuclear Research (CERN) in recent years to conduct fundamental nuclear physics research, particularly studying the basic constituents and interactions of the microscopic world. CERN's research focuses include particle physics, string theory, quantum field theory, relativity, and cosmology. CERN houses the world's largest particle physics experimental facilities, including two main laboratories: the Large Hadron Collider (LHC) in Switzerland and the Super Proton Synchrotron (SPS) in France. The LHC, currently the largest and most powerful particle accelerator in the world, mainly aims to simulate the initial state of the universe after the Big Bang, explore the interactions between fundamental particles, and investigate new physical phenomena and unknown particles. The SPS, a smaller accelerator, is mainly used to provide high-energy and strong-beam flow for the LHC. CERN's achievements in particle physics research are outstanding, such as discovering the W and Z bosons, hadrons, quarks, and antimatter. The most famous one is the discovery of the Higgs boson in 2012, which is a massive fundamental particle and was a landmark event in modern physics.

4.1.2.2 *Military field*

4.1.2.2.1 Nuclear weapon development in Germany

During World War II, Germany used to set a dedicated military application office for nuclear fission, which planned to conduct chain fission reactions using uranium and graphite and to study methods for separating uranium isotopes. Some measurement results for graphite were later proven to be incorrect, leading physicist Werner Heisenberg to suggest using heavy water instead of graphite as the moderator material. As the only heavy water production facility in Germany was located in Norway and could be easily targeted and destroyed by the Allies, the German team did not have the industrial support they needed. This dependence on scarce heavy water was a major reason why the German experiment never succeeded.

4.1.2.2.2 Nuclear weapon development in the former Soviet Union

The former Soviet Union took advantage of information obtained from the "Manhattan Project" of the United States and used plutonium instead of uranium as the main material for atomic bombs. In December 1946 the former Soviet Union's first graphite-moderated reactor was launched in southern Russia to achieve controlled chain fission of uranium. Autumn 1948 saw the commissioning of the first plutonium-producing reactor in the former Soviet Union. On August 29, 1949 the former Soviet Union successfully detonated its first atomic bomb, becoming the second country to own nuclear weapons in the world. On November 22, 1955 the former Soviet Union successfully conducted its first air-dropped hydrogen bomb test, with an estimated yield of 1.6 million tons. This is marked as the first time that hydrogen bombs were capable of being used in actual combat. During the period of the mid-1940s to early 1990s the former Soviet Union developed a wide variety of nuclear weapons, including atomic and hydrogen bombs, neutron bombs, and so on, as well as deep exploration in the delivery systems such as bombers, long-range ballistic missiles, and nuclear submarines. The power of nuclear weapons was multiplied by thousands.

4.1.2.2.3 Nuclear weapon development in Russia

In the 1990s, after the dissolution of the former Soviet Union, Russia inherited the majority of former Soviet Union nuclear weapons and modernized its strategic nuclear capabilities. This included everything from strategic nuclear weapons delivery vehicles to launch platforms and new nuclear-capable nuclear-powered "Poseidon" torpedoes. At the same time, Russia also sought to enhance its strategic and tactical nuclear weapons deterrence and combat capabilities by deploying multiple nuclear-tipped and conventional-tipped hypersonic vehicles. With its power greatly reduced compared to the former Soviet Union era, Russia relies even more on nuclear weapons to maintain its status as a major power. For example, Russia deployed the second-generation dual-capable "Oreshnik" intermediate-range ballistic missiles during the Russia–Ukraine war, with ranges of several thousand kilometers. In October 2025 it also test-fired "Petrel" nuclear-powered cruise missile and has successfully developed the "Poseidon" nuclear-powered underwater unmanned vehicle.

4.1.2.2.4 Nuclear weapon development in the United Kingdom

On October 3, 1952 the United Kingdom successfully developed its own nuclear weapons program under the leadership of Chadwick, who had participated in the "Manhattan Project." They detonated a nuclear weapon underwater at Monte Bello Islands in Australia, becoming the third country to own nuclear weapons. On May 17, 1957 the United Kingdom also conducted its first hydrogen bomb test. The United Kingdom subsequently established a strategic nuclear force in the sea, land, and air [2]. However, because of the United Kingdom's reliance on US nuclear protection, it pursued a minimum nuclear deterrence strategy. After the end of the period of the mid-1940s to early 1990s the United Kingdom analyzed the international security environment and threats and decided to maintain only its sea-based strategic nuclear forces in 1998. After 2023 the United Kingdom has continued advancing construction of the "Dreadnought" class strategic nuclear submarines, expected to enter service in the 2030s, and continues policy debates over updates to the "Trident" missile system and the scale of nuclear warhead stockpiles, maintaining its sea-based nuclear deterrent.

4.1.2.2.5 Nuclear weapon development in France

On February 13, 1960 France successfully developed its own nuclear weapons under the leadership of Joliot-Curie. They conducted their first atomic bomb test at Reggane in the Sahara Desert in French West Africa with an estimated yield of about 60,000 t, becoming the fourth country to own nuclear weapons. On August 24, 1968 France successfully conducted a short-range test of the hydrogen bomb [2]. France has always pursued an independent nuclear strategy and attached importance to the role and status of nuclear weapons. France's nuclear force has always ranked third in the world, with a maximum deployment of 540 nuclear warheads during the period of the mid-1940s to early 1990s and currently has no more than 300 nuclear warheads. In 1996 France abandoned its land-based nuclear forces and retained only its air- and sea-based strategic nuclear forces. The future nuclear arsenal of France will mainly consist of four "Triomphant" class ballistic missile submarines and a small number of air-based nuclear warheads. After 2023 France accelerated development of the third-generation strategic nuclear submarines (SNLE 3G), planned to gradually replace the current "Triomphant" class in the 2030s, and continued deployment of the new air-launched cruise missile (ASMP-A) and its successor systems.

4.1.2.2.6 Deployment of US nuclear weapons in Europe

Europe has been playing an important role in the US global defense strategy for a long time. The deployment of nuclear weapons in Europe is also an important component of the US global nuclear strategy. During the Russia–Ukraine conflict the United States accelerated its deployment of more advanced tactical nuclear weapons in European bases and strengthened its nuclear combat capability and control over Europe.

4.1.2.2.7 Evaluation

Nuclear weapons are one of the most destructive weapons in the world. Possessing nuclear weapons means having considerable international influence and deterrent power. After World War II, European countries accelerated the research and development of nuclear weapons to break the US nuclear monopoly. However, as the international situation undergoes transformation and tension intensifies, the United States tends to deploy more nuclear weapons in Europe.

4.1.2.3 Energy and power field

4.1.2.3.1 Nuclear power development in the United Kingdom

Because of the fact that the United States practically monopolizes the uranium enrichment technology in the Western world, the United Kingdom has adopted different strategies for the development of nuclear power and has produced a series of natural uranium-based, graphite-moderated, and gas-cooled reactors. The first 50 MW Magnox reactor began operation in 1956 and remained in operation until 2003. From 1965 to 1988 some advanced gas-cooled reactors were established, and all of them will be closed by 2030. The Sizewell B nuclear power station was the only one built in Suffolk in 1995. In the United Kingdom, from 2000 to 2021, the annual electricity share provided by nuclear power decreased from 23.0% to 14.9%. From 2000 to 2021 the share of electricity generated by nuclear power in the United Kingdom declined from 23.0% to 14.9%. In 2024 and 2025 nuclear power accounted for roughly 14%–15% of the country's electricity generation. In 2024 the UK government launched a nuclear power expansion plan, aiming to triple nuclear capacity by 2050. The Hinkley Point C nuclear power project is a joint project among China, France, and the United Kingdom. It was originally financed and constructed by a Chinese consortium led by China General Nuclear (CGN) together with EDF Group, using the EPR technology. After CGN stopped providing funding for Hinkley Point in 2023 the project has been led by EDF, with continued support from the UK government, whereas United States private equity capital has entered to help fill the financing gap. In November 2025 the UK government formally approved the final investment decision for the Sizewell C nuclear project, which adopts the same EPR technology as Hinkley Point C, marking the substantive advancement of another large-scale nuclear project.

4.1.2.3.2 Nuclear power development in France

France began to adopt a gas-cooled graphite-moderated reactor design similar to Magnox, with the first reactor starting operation in 1956. In 1963 the first commercial reactor, Chinon A-1, was connected to the grid. In 1973, because of the oil crisis and embargo, France had no coal or natural gas, and nuclear energy became the preferred energy source, adopting the pressurized water reactor technology route. The 1980s was the decade that nuclear power experienced the fastest growth in France, with multiple reactors being built. Currently, France operates 56 second-generation pressurized water reactors and one EPR reactor under construction. In 2024 the EPR reactor was officially connected to the grid and began generating electricity. Approximately 72% of France's electricity is produced from nuclear energy, with a total installed capacity of 61.37 GWe, ranking second in the world. In 2025 the French government is actively promoting lifetime extensions and modernization upgrades for the existing 56 reactors and has formally launched a program to build at least 6 and up to 14 next-generation EPR-2 reactors by 2035, with the first unit targeted to enter operation in 2035.

4.1.2.3.3 Nuclear power development in the former Soviet Union

In 1954 the former Soviet Union built its first nuclear power plant, the 5 MW Obninsk reactor, which was also the world's first nuclear power reactor for electricity generation. In 1964 the first two commercial nuclear power reactors of the former Soviet Union were launched, including a 100 MW boiling water graphite-moderated reactor (RBMK) and a 210 MW pressurized water reactor (VVER). Subsequently, both reactors were converted to a 1000 MW standard design and became two major nuclear power reactor types in the former Soviet Union. By the mid-20th century, there were 25 power reactors operating in the former Soviet Union [3]. From the Chernobyl accident in 1986 to the mid-1990s, only one nuclear power plant was in operation in the former Soviet Union.

4.1.2.3.4 Nuclear power development in Russia

After the dissolution of the Soviet Union, from the early 1990s to the late 1990s, Russia negotiated reactor export agreements with Iran, China, and India, while gradually resuming stalled domestic nuclear construction within the limits of available funding. These negotiations led to reactor export contracts for the Bushehr nuclear power plant in Iran, Tianwan nuclear power plant in China, and Kudankulam nuclear power plant in India [3]. The deals provided Russia's nuclear industry with crucial financial support and international credibility, enabling it to survive the difficult post-Soviet period and laying the foundation for later revival. Around 2000, nuclear construction resumed domestically; Rostov Unit 1 began operation in 2001, greatly boosting morale within the Russian nuclear sector. It was followed by Kalinin Unit 3 in 2004, Rostov Unit 2 in 2010, and Kalinin Unit 4 in 2011. In 2017 the first VVER-1200 nuclear power unit began commercial operation, representing a major advancement over the VVER-1000 design, with significant improvements in both economic efficiency and safety. By 2025 Russia operates 37 nuclear power units, with nuclear power accounting for over 19% of total electricity generation. Russia's long-term strategy is to transition toward inherently safe fast reactors using a closed nuclear fuel cycle by 2050, an area where the country maintains relatively advanced technological capabilities. In 2025 key construction milestones were reached for Unit 2 of Leningrad II and Unit 2 of Kursk II nuclear power projects.

4.1.2.3.5 Nuclear power development in Germany

In Germany the relationship between nuclear power and public acceptance has been a subject of contention for a long time, with both a certain scientific and political interest in nuclear energy and limited social acceptance of it. In June 1961 the Kahl experimental nuclear power plant was connected to the grid for the first time in the Federal Republic of Germany. Five years later, the Reinsburg nuclear power plant became the first commercial nuclear power plant in the German Democratic Republic. In the early years, Germany built various types of reactors, including heavy water-cooled pressurized water reactors (MZFR in Karlsruhe), pebble bed reactors (AVR), pressure tube reactors (Niederaichbach nuclear power plant), superheated steam reactors (HDR in Karlstein), and sodium-cooled breeder reactors (SNR-300 in Kalkar). Because of various reasons, Germany did not further promote the above reactor concepts. Because of the oil crisis from 1973 to 1985, Germany built 40 nuclear reactors, mainly light water reactors. In the 1990s the share of nuclear energy in the German power structure was around 30%, but by 2022, this proportion was only 6.5%, with only three nuclear power plants still in operation. Since the Fukushima accident in 2011 the federal government, supported by a broad consensus in society, eventually decided to phase out nuclear energy, and since then, nuclear power generation has been steadily declining. On April 15, 2023 the last nuclear power plant in Germany was also shut down. In May 2025 Germany stated that it would no longer oppose France's efforts within European Union (EU) legislation to grant nuclear energy the same status as renewable energy. Germany's support is expected to provide new momentum for the development of nuclear power technology in Europe. In 2025 the German government also approved an action plan titled "Germany's Path Toward a Nuclear Fusion Power Plant" at a federal cabinet meeting in Berlin, aiming to build the world's first nuclear fusion power plant in Germany.

4.1.2.3.6 Nuclear power development in other European countries

A total of 13 of the 27 EU member states have a total of 103 reactors in operation. There are different views on the use of nuclear energy throughout the EU. Because of the Russia–Ukraine conflict the European continent is struggling to cope with the energy crisis. Some EU member states fully support nuclear energy because nuclear reactors meet most of their energy needs, whereas some other member states have a complex history of security concerns and negative public opinion on nuclear, which makes nuclear energy completely excluded from their energy system. After the Fukushima nuclear accident, Germany permanently closed eight reactors. Switzerland and Spain have banned the construction of new reactors. Belgium is considering phasing out nuclear power plants. By 2030, because of the closure of some reactors, it is expected that the nuclear capacity of the EU will slightly decrease in the near future. However, there are also EU member states, such as France and Slovakia, that are currently building new nuclear power plants. In 2023 countries such as Sweden and the Netherlands clarified their plans to build new nuclear power plants. In 2024 Poland's first nuclear power plant project entered a critical approval stage. In 2025 the Swedish lead-cooled small modular reactor (SMR) company Blykalla signed a memorandum of understanding with the Swedish National Research Institute to jointly accelerate the development of advanced reactors.

4.1.2.3.7 Nuclear fusion development in Europe

By the 1950s researchers began to study the possibility of replicating the nuclear fusion process on Earth. In 1950 Andrei Sakharov and Igor Tamm, scientists from the former Soviet Union, proposed a design for a magnetic confinement fusion device, known as the Tokamak. The currently built Tokamaks in Europe include the Joint European Torus (JET) and the Mega Amp Spherical Tokamak (MAST). The ITER project currently under construction in Cadarache, France, will be put into operation in the 2020s and will become the largest Tokamak [4,5]. Europe's Tokamak facilities include the JET and the MAST, which are jointly operated by multiple European countries.

In 2020 the major installation phase of the ITER project officially began with a ceremony held at the organization's headquarters in Cadarache, France. The ITER project has now entered a 5-year assembly phase, with an initial plan to conduct the first plasma test in December 2025, though this has since been postponed to 2033. In 2025 several key components achieved progress. For example, Europe completed the manufacture of the second vacuum vessel sector, and all cryopumps have been delivered and transported to the ITER site.

4.1.2.3.8 Evaluation

After World War II, nuclear energy transitioned from military use to civilian use, particularly in the area of electricity generation. Countries such as the United Kingdom, France, Germany, and Russia started developing nuclear power technology at an early stage at that time. At present, Germany has shifted from a nuclear phase-out stance to no longer obstructing the development of nuclear power. Meanwhile, the United Kingdom has launched a nuclear expansion plan, and European countries such as France and Russia are continuing to construct new nuclear power projects.

4.1.2.4 Nuclear technology field

4.1.2.4.1 Applications in industry

Nuclear technology is widely used in industrial fields in Europe, including e-beam melting (EBM), semiconductor industry, neutron imaging, and so on. Among these industry fields, EBM technology can produce highly complex metal parts, and the pioneer of this technology is Arcam, a Swedish company recently acquired by General Electric. Ionizing-radiation tools are used in the manufacture of semiconductor integrated circuits (IC), using ion-implant equipment and process control, and also in silicon-wafer production. Commercially available accelerator systems for ion implantation are offered by various vendors in Europe, such as Ion Beam Services (France), a pioneering company specializing in plasma immersion ion implantation (PIII) and SiC implanters, as well as high voltage engineering (Netherlands), along with multiple radio-frequency (RF) linear accelerator providers. Neutron computed tomography (CT) allows the visualization of the inner structure of the sample. Numerous university research reactors and national laboratories possess advanced capabilities for neutron imaging, such as ORPHEE in France and FRM II in Germany.

4.1.2.4.2 Applications in agriculture

In Europe, nuclear technology has several concrete applications in agriculture. These include food irradiation, which is used to eliminate pathogens, extend shelf life, and serve as an alternative to certain pesticides; radiation-induced mutation breeding, which helps develop crop varieties that are more resistant to environmental stress and better adapted to climate change; radioactive tracers, used to measure and monitor soil moisture and nutrient flows to optimize irrigation and fertilization; and stable isotopes, which are applied to study soil–plant interactions and evaluate nutrient use efficiency, thereby improving soil management and increasing crop yields.

4.1.2.4.3 Applications in the medical industry

The medical use of radiation and nuclear technology has benefited patients throughout Europe. The EU is a global leader in the supply of medical radioisotopes and the development of radiological diagnosis and treatment. In addition, it has taken the lead in developing legal standards for quality and safety in radiology, radiotherapy, and nuclear medicine. Nuclear medicine uses radioactive substances, mainly for the diagnosis of cancer, heart disease, and other diseases. At the end of the 19th century, the discovery of nonenergy radiation in the medical field quickly promoted its wide applications in health imaging and therapy. According to data from the medical industry, 500 million EU citizens receive more than one radiological examination every year on average, including dental screening, chest imaging, breast cancer screening, or CT. The EU performs more than 55 million CT procedures annually, as well as approximately 10 million nuclear medicine imaging procedures, including single photon emission CT (SPECT) and positron emission tomography (PET), occasionally combined with CT or magnetic resonance imaging (MRI).

4.1.2.4.4 Applications in public security

With the complex and changing international situation, modern nuclear technology plays an increasingly important role in preventing and combating various terrorist and criminal activities, such as using X-ray, accelerator, and radiation source imaging technology for container, vehicle, and luggage inspection. The application industry in Europe is mature and is at an international advanced level.

4.1.2.4.5 Evaluation

Nuclear technology is a powerful tool in the industrial field to support the development of all six key technologies. In the medical field, it is used for radiological diagnosis and treatment. In the agricultural field, it is used for sterilization, insect control, breeding, monitoring water and nutrients, etc. In the public security field, it is used to prevent and combat various terrorist activities and criminal activities, providing great security for the people of Europe in terms of clothing, food, shelter, and transportation.

4.1.3 Impact factors on development of European nuclear science and technology

4.1.3.1 Economic development factors

The development of nuclear technology can not only reduce the dependence on imported energy and provide competitive power production for the economic development of Europe but also create markets for equipment, metallurgy and construction industries, and electronics, as well as numerous commercial service sectors, including finance and insurance, bringing extensive economic benefits to the whole Europe, maintaining millions of jobs, and increasing several trillion euros of Gross Domestic Product (GDP) growth and tax revenue.

4.1.3.2 Environmental protection factors

As a proven low-carbon emission, clean, and stable baseload power source with high energy security, nuclear energy provides one-quarter of the EU's carbon-free electricity, whereas there are different views on the use of nuclear energy across the EU. Advocates of nuclear power development highlight several environmental advantages, including the emission reduction of smog and other climate benefits resulting from its low greenhouse gas emissions. Among the opponents to nuclear power, concerns focus on the challenges associated with safe waste disposal, and the perceived elevated likelihood of accidents resulting in radioactive contamination is prominent.

4.1.3.3 Science and technology factors

The member states of CERN, established in Europe, have the same scientific and technological blueprint, making it easier for them to engage in harmonious scientific and technological cooperation and sharing. At the same time the research results of nuclear science and technology at CERN can raise human understanding of the material world to a new level, laying a good foundation for the development of nuclear science and technology in Europe.

4.1.3.4 Social development and public psychology factors

European countries have promoted the peaceful use of nuclear energy, which can not only provide green energy but also improve people's quality of life, protect people's health and safety, and promote social and economic growth. However, because of the impact of major nuclear accidents such as Chernobyl, Three Mile Island, and Fukushima, the public has also developed fears of nuclear energy.

4.1.3.5 National policy and governance factors

Within the EU the peaceful use of nuclear energy is governed by the 1957 Treaty establishing the European Atomic Energy Community (Euratom), which provided a legal framework for its authority and activities. Euratom sets requirements for nuclear safety, radioactive waste, and spent fuel, as well as basic safety standards. Europe has developed a set of effective policies and management systems for nuclear energy management.

4.1.3.6 Military factors

During World War II and the period of the mid-1940s to early 1990s, weapons became the most needed thing for the majority of European countries at that time. Under this special historical background, major European powers actively developed nuclear weapons for political needs. At present, nuclear weapons are regarded as a deterrent military force by some countries.

4.2 Impact of nuclear science and technology on Asia

4.2.1 General development of Asian nuclear science and technology

4.2.1.1 Development status

Among the Asian countries, China, Japan, and South Korea are the most important driving countries for the development of nuclear science and technology. The industries of nuclear technology applications in the three countries have a solid foundation, extensive business, and outstanding characteristics. Japan holds technological advantages in nuclear reactors, nuclear medicine, and industrial nuclear applications and has developed a mature system for research, development, and administrative management. South Korea has a wide range of nuclear technology applications and ranks among the world's leaders in the number of technical equipment and products per capita; its indigenously designed APR-1400 pressurized water reactor has already achieved successful international exports. China is currently the fastest-growing country in the world in terms of nuclear technology applications, and nuclear science and technology have been elevated to a strategic emerging industry strongly supported at the national level. Nuclear power generation is playing an increasingly important role in China's energy development, whereas nuclear technologies are also widely used in industry, agriculture, and medicine. India has also constructed nuclear power plants and actively promotes nuclear technology development. Meanwhile, an increasing number of Southeast Asian countries are paying attention to the development of nuclear energy.

In addition, Asian countries represented by China, Japan, and South Korea attach great importance to the development of nuclear fusion. China operates fusion devices such as the Experimental Advanced Superconducting Tokamak (EAST) and the HL-3 device. Japan also hosts many fusion experimental facilities, including the large Tokamak device JT-60SA and the large helical device (LHD). On December 25, 2025, Japan's National Institute for Fusion Science (NIFS) officially announced that the world's first large fully superconducting helical device, after completing its 26th round of experiments, had ceased operation, marking the successful conclusion of a 27-year physics research program. South Korea is also among the world's leading countries in nuclear fusion research, with facilities such as the Korea Superconducting Tokamak Advanced Research (KSTAR) device.

4.2.1.2 Development trends

Compared with Western countries the development of nuclear science and technology in Asia started relatively late, but it has developed rapidly and has been widely applied in industry, agriculture, medical industry, environmental protection, and other fields. It has formed an industrial scale in irradiation material modification, irradiation processing services, radiation technology equipment, public health, public safety, environmental protection, and other fields and has achieved significant economic and social benefits. In areas such as the design and application of research reactors and SMRs as well as other advanced nuclear energy systems, Asian countries are actively making strategic deployments, providing key platforms for the research, development, and industrialization of nuclear science and technology. In the future, nuclear science and technology will be deeply integrated with other technologies, industries, and various fields of economic and social development, promoting the progress and development of Asian countries in various aspects.

4.2.2 Specific development of Asian nuclear science and technology

4.2.2.1 Scientific research field

Nuclear science and technology in Asia, in particular atomic nuclear physics, have developed rapidly with the development of accelerator technology and detection technology. At present, fundamental atomic nuclear physics, including the excitation energy and density of nuclei, angular momentum, isospin, and mass degrees of freedom, has continued to open new frontiers at its extremes. At the same time a series of new phenomena have been discovered, which have led to fundamental challenges and breakthroughs in atomic nuclear physics. The construction of large-scale nuclear physics experimental devices and the establishment of quantum chromodynamic theory have greatly expanded the field of atomic nuclear theory research.

Asian countries have put forward different development plans based on their research foundation and dominant position in the field of nuclear science as well as the major needs of national development, including high energy physics, cosmic rays and high energy astrophysics, nuclear physics (heavy ion physics), plasma physics, theoretical physics, nuclear technology scientific detectors and accelerators, synchrotron radiation, and free electron lasers. The development of nuclear technology will focus on new principles, methods, and applications of nuclear analysis, including research on several fundamental issues in ion beam and radiation technology science. Based on the single-molecule physics, biology, and molecular nuclear medicine of nuclear technology and nanotechnology, research on radiopharmaceuticals and particle beam therapy technology is carried out.

4.2.2.2 Military field

4.2.2.2.1 Nuclear weapon development in China

China is the only major country in Asia that legally possesses nuclear weapons. From 1946 to 1953 China began to prepare investigations and research on nuclear weapons. During this period, it jointly explored uranium mines in Xinjiang with the Soviet Union and established the Atomic Energy Commission. From 1955 to 1959 China's nuclear weapon research developed rapidly. At that time, China made multiple cooperation agreements with the Soviet Union. After 1959 China embarked on the road of self-reliance. It successfully detonated the atomic bomb and hydrogen bomb in 1964 and 1967, respectively. After 1974 China began to establish a strategic missile combat system. In 1992 China signed the Treaty on the Non-Proliferation of Nuclear Weapons (NPT). At present, China's nuclear strategy is the minimum effective nuclear deterrence. Under the guidance of this strategy, China has also placed emphasis on the development of nuclear warheads and intercontinental ballistic missiles (ICBMs).

4.2.2.2.2 Nuclear weapon development in the remaining Asian countries

The other Asian countries are not on the list of countries that are able to legally possess nuclear weapons. However, India, Pakistan, and the Democratic People's Republic of Korea have all conducted de facto open nuclear tests. In terms of the quality of nuclear weapons, all these countries have Generation I nuclear weapons, i.e., atomic bombs, whereas no country owns more powerful hydrogen bombs. However, periodic calls in Japan and South Korea for the acquisition of nuclear weapons are regarded as a dangerous signal. In October 2025, during a visit to South Korea, US President Donald Trump stated that the United States could support South Korea in building nuclear submarines in the United States.

4.2.2.2.3 Evaluation

Nuclear weapons are a major kind of weapon in national military strategy. As a guarantee of national security, countries that possess nuclear weapons tend to have greater overall strength. How nuclear weapons are utilized and how effective nuclear deterrence is carried out has a bearing on the fundamental security of all countries in international relations. With the development of nuclear weapons, it has been realized that it is necessary to limit the proliferation of nuclear weapons and refrain from adding new ones or developing them.

4.2.2.3 Energy and power field

4.2.2.3.1 Nuclear power development in China

China began to establish its nuclear industry in the 1950s. But the real history of nuclear power in China began in the early 1970s. In 1985 China started the construction of the first nuclear power plant of its own design, the Qinshan nuclear power plant, which was connected to the grid for the first time in December 1991, ending China's history of lack of nuclear power. In February 1994 and May 1994 two sets of M310 MW nuclear power units imported from France were put into commercial operation in Daya Bay, Guangdong Province, marking China's first international cooperation in the field of nuclear power construction. Through the introduction, digestion, and absorption of AP1000 technology, China successfully developed the third-generation passive nuclear power design "Guohe One," which possesses independent intellectual property rights. In January 2021 the Hualong One nuclear power unit, a Generation III pressurized water reactor with China's fully independent intellectual property rights, was put into commercial operation. After decades of development, China has now established a comprehensive nuclear industry system. From uranium enrichment technology and nuclear fuel manufacturing to the development of nuclear power plant technology, spent fuel reprocessing technology, and research on fast reactors, such as high-temperature gas-cooled reactors and small reactors, as well as fusion reactors. These technologies are at the forefront of the world. In 2024 several nuclear power units using the "Hualong One" technology—such as Unit 1 of the Taipingling nuclear power plant in Guangdong and Unit 1 of the Jinqimen nuclear power plant in Zhejiang—were successively put into commercial operation. In September 2025 the world's first land-based commercial modular small reactor, "Linglong One," completed the installation of its core module and containment vessel tests at Changjiang, Hainan, and also completed its cold functional tests.

4.2.2.3.2 Nuclear power development in Japan

Japan is a country that relies heavily on imported energy, with more than 80% of its energy being imported. After World War II the government of Japan decided to develop nuclear and to utilize it peacefully for its economic development.

Since 1954, Japan has made heavy effort in the investigation and development of its nuclear industry. By 1985, Japan had built 15 nuclear power plants and more than 30 nuclear power units with a total installed capacity of more than 23,000 MW. More than a decade thereafter the development of nuclear power industry in Japan continued to accelerate. At the end of the last century, Japan built a total of 51 nuclear power plants. Japan has also built the world's first Generation III boiling water reactor and became the first to put it in operation. Unfortunately the Fukushima nuclear accident occurred during the rapid development, leading to difficulties in nuclear industry development in Japan. However, as nuclear energy is still an inevitable choice to ensure energy supply, especially in the process of global clean energy transformation. In response to these developments, the Japanese government released the "Fifth Strategic Energy Plan" in 2018, continuing to promote the restart of nuclear power plants under the premise of safety [6,7]. In 2025 the Japanese government and power companies actively promoted approvals for lifetime extensions of existing nuclear power units and explored the possibility of constructing next-generation reactors. However, on January 21, 2026, Unit 6 of the Kashiwazaki-Kariwa Nuclear Power Plant, which had just restarted, issued an alarm during operations on January 22. Because of the time required for investigation, authorities decided to temporarily shut down the unit.

4.2.2.3.3 Nuclear power development in South Korea

The Korean government established the Atomic Energy Department in 1956 and became a member of the International Atomic Energy Agency (IAEA) in 1957. In the mid-1980s the Korean nuclear industry began to standardize the design of nuclear power plants and to cooperate with Western companies. Korea has built several nuclear power plants, including Guri 1 and Guri 2, and the Wolsung nuclear power plant, through self-construction and cooperative construction. With the Fukushima accident and its problems leading to anxiety about nuclear energy among the Korean public, the government began to shut down many nuclear power plants. However, as South Korea's way to develop other energy sources is very hindered, to ensure energy security, the South Korean government began to reexamine the development of nuclear power. It will restart nuclear power units and promote the growth of installed nuclear power capacity in operation to about 4.8 GW. South Korea is expected to complete the construction of Units 3 and 4 of the Shin-Kori nuclear power plant by the end of 2025 and promote their full-power operation while continuing the long-term operation plans for existing reactors. These efforts aim to achieve the government's goal of increasing the share of nuclear power to more than 30% of the national energy mix.

4.2.2.3.4 Nuclear power development in India

India has one of the most favorable locations in the South Asian subcontinent, but it has a shortage of fossil energy. India is closely linked to Western countries in the nuclear power development field. In 1948 the Atomic Energy Commission of India was set up, and the Department of Atomic Energy was formed to be responsible for the research and development of nuclear technology and the operation of commercial reactors. In 1974 India detonated a nuclear device, which led to the discontinuation of civil nuclear energy cooperation with most of the countries in the world. This has resulted in India having only limited access to nuclear technology and materials from the international market, which has severely constrained the development of India's nuclear power business. With the profound changes in the international political landscape, the international community has lifted civil nuclear sanctions against India. In 2025 India's nuclear power development entered a phase of rapid expansion. Owing to the enormous market value potential of India's nuclear energy sector, countries such as France, Russia, South Korea, Argentina, Canada, and the United Kingdom have successively signed civil nuclear cooperation agreements with India.

4.2.2.3.5 Nuclear power development in the remaining Asian countries

By the end of 2021, 25 out of 47 Asian countries had made carbon neutral commitments. This has led Asian countries that do not have nuclear power generation capacity to plan or consider building nuclear power plants. Pakistan and Iran have already built nuclear power plants. Bangladesh, Indonesia, Jordan, Kazakhstan, Malaysia, Saudi Arabia, Thailand, Turkey, North Korea, Israel, and Vietnam have plans or are considering building nuclear power plants. Currently, Asia is continuing to lead the global growth in installed nuclear power. Asia accounts for 18% of global nuclear power generation.

4.2.2.3.6 Evaluation

Energy is a problem that cannot be avoided in the development of Asia and even the whole world. At the same time, as people gradually raise their awareness of environmental protection, the demand for clean energy is also gradually rising. Energy structure is the lifeblood of a country's economic development. A reasonable energy structure can avoid the impact of potential risks as much as possible. Carbon neutrality is the road to sustainable development, and the use of nuclear energy can effectively reduce carbon emissions and promote the upgrading of energy and related industries. Nuclear, wind, and photovoltaic energy storage is the blueprint for development. Nuclear energy and its combined hydrogen production are also developing.

4.2.2.4 Nuclear technology field

4.2.2.4.1 Applications in industry

Industrial applications of nuclear technology in Asia are developing rapidly and now involve the whole industry chain, forming a huge market. At present the main applications are industrial radiography, nondestructive testing, nuclear measurement, and material modification. China has built about 130 cobalt source irradiators, and the number of accelerators for various types of irradiation applications is growing rapidly. Detectors in the public security field are also developing. Nondestructive testing of gas pedals from Japan and South Korea is widely used in aerospace, aviation, nuclear power, machinery manufacturing, and other fields.

4.2.2.4.2 Applications in agriculture

According to a United Nations survey, 55% of the population of the world lives in Asia. With such a large population and density the demand for food is very large. Asia accounts for one-third of the world's arable land and is highly intensified and fragmented [8]. Therefore, finding crop seeds that are suitable for most kinds of soils, having higher yields and adapting to various extreme conditions such as dryness and water shortage, is an important measure to solve the contradiction between development and soil.

Mutation breeding has been used as early as the 1930s to accelerate the acquisition of superior plant traits. This method is based on the plant's genome. It uses physical radiation or chemical means to induce spontaneous genetic variation in the plant. The mutant imitates the natural mutation process and produces random genetic mutations to produce the new useful traits needed [9]. Currently, irradiation mutation breeding is a mature technology that is less expensive, faster, more universally applicable, harmless, and environmentally friendly compared to other methods. According to Sobhana Sivasankar, Chief of the Plant Breeding and Genetics Section of the Joint FAO/IAEA Centre of Nuclear Techniques in Food and Agriculture, among the recorded crop varieties in Asia and the Pacific, radiation mutation breeding crops account for nearly 2000. Among the 826 rice varieties using radiation mutagenesis, 699 originated from Asia and the Pacific. In addition, as an important means of food safety, irradiation sterilization has been widely used in Asia, contributing greatly to the Asian food crisis and food safety.

4.2.2.4.3 Applications in the medical industry

Compared to other nuclear application fields, nuclear medicine in Asia started relatively late, but it develops rapidly. In the nuclear medicine area, China and Japan are the most prominent pioneer countries in Asia. In Japan the application of nuclear medicine has increased dramatically. There has been a slight increase in the use of CT/PET for cancer diagnosis and particle beam therapy. China has released the medium- and long-term development plan for medical isotopes (2021–35), which incorporates nuclear medicine development into the country's future development plan. Nuclear neutron, proton, and heavy ion therapy instruments are also booming in China.

4.2.2.4.4 Evaluation

With a large population, solving the food problem in Asia is solving the survival issue. Asia is a vast area covering a wide variety of soil, terrain, and climate. The potential and scale of nuclear technology applications in industry are remarkable and will be one of the focuses of future development. Radiation mutation breeding can effectively breed crops suitable for different requirements and provide effective measures to solve the food crisis. Nuclear medicine can provide clinicians with unique insights into the cellular and molecular processes of patients, promoting the completion of early diagnosis and treatment. At the same time the emergence of various radiopharmaceuticals opens more possibilities for oncology.

4.2.3 Impact factors on development of Asian nuclear science and technology

4.2.3.1 Economic development factors

As the energy transformation continues, Asian countries that lack a stable energy source to replace traditional energy sources have been facing difficulties. For some countries in Asia, nuclear energy is still the inevitable choice for securing energy supply. In the long term, accelerating the construction of new nuclear power plants and the application of nuclear technology will also contribute to sustainable economic growth.

4.2.3.2 Environmental protection factors

With the goal of carbon neutrality, nuclear energy as a clean and low-carbon emission energy source makes it irreplaceable in the global clean energy transformation. Therefore many Asian countries still consider nuclear energy development one of their key priorities.

4.2.3.3 Science and technology factors

To promote nuclear science and technology in Asia the Asia Pacific Regional Nuclear Cooperation Agreement was established in 1972 with the support of the IAEA. It aims to promote, coordinate, and implement cooperative research, development, and training projects in the peaceful use of nuclear science and technology among its members. Cooperation is enhanced through projects focusing on the common specific needs of members to promote and coordinate the nuclear science and technology development.

4.2.3.4 Social development and public psychology factors

Because of the social development issues faced by most of the Asian countries, as well as the various development levels of science and technology, the nuclear science and technology development is mainly concentrated in China, Japan, South Korea, and India. At the same time the nuclear science and technology in Asian countries are also constrained by the sustainable supply of nuclides. Moreover, there are still deficiencies in the mechanism for guaranteeing the supply of natural uranium [6]. As a result of nuclear accidents such as Chornobyl, Three Mile Island, and Fukushima [7], the public has a certain degree of fear of nuclear.

4.2.3.5 National policy and governance factors

Before the Fukushima nuclear accident, policies in Asian countries were very active in nuclear power development. South Korea cooperated closely with Western countries and actively marketed its nuclear power technology. China planned a large number of nuclear units under construction. Iran expanded its uranium production and proposed to increase its nuclear power generation capacity without relying on international partners. Overall, Asian countries strongly support nuclear power development and see it as a key to transformation. The Fukushima nuclear accident has led to increased opposition to nuclear energy development. In the decade following it, 12 nuclear reactors permanently shut down and 24 reactors decommission in Japan. South Korea and other countries also shut down numerous nuclear power units after the accident. Currently, Asian countries are beginning to restart nuclear power in an orderly manner with safety. China has also developed additional laws, regulations, and standards to continue nuclear power development. Overall the Fukushima nuclear accident has made Asian countries more cautious about developing nuclear power, paying more attention to nuclear safety, and adopting more technologies, management, laws, and regulations. With the global emergence of the goals of "carbon peak" and "carbon neutrality," nuclear power—recognized as a stable, reliable, and low-carbon energy source—has seen its strategic value reassessed. Driven by the dual imperatives of ensuring energy security and addressing climate change, many Asian countries have incorporated nuclear energy as a core component of their national low-carbon energy systems, initiating a new round of accelerated development to support their green economic and social transformation.

4.2.3.6 Military factors

After World War II, Asian countries began to develop nuclear weapons to enhance their national defense strength. Some countries have also promoted military nuclear power, such as nuclear submarines, to circumvent the restrictions imposed by the NPT.

4.3 Impact of nuclear science and technology on North America

4.3.1 General development of North American nuclear science and technology

4.3.1.1 Development status

In North America the United States and Canada lead the world in nuclear science and technology, both in military and civilian applications. The United States detonated the world's first atomic bomb. In the nuclear power field the United States designed the first fully commercial pressurized water reactor and developed boiling water reactors. By the late 1960s the United States had developed pressurized water reactors and boiling water reactors with capacities exceeding

1000 MW. After the Three Mile Island accident in 1979 the US nuclear industry faced significant setbacks and entered a period of stagnation that lasted for 20 years. In 2019 the US nuclear energy production had increased to 809 TWh, accounting for nearly 20% of the US electricity generation and over 30% of global nuclear energy production. Canada has over 60 years of experience in the nuclear energy field. It involves uranium mining, nuclear power plant design, construction and operation, decommissioning, waste management, regulatory oversight, and governance. Canada is known for developing the unique Canada Deuterium Uranium (CANDU) nuclear reactor. In addition, Canada boasts a robust nuclear supply chain that serves the CANDU reactors both domestically and internationally.

4.3.1.2 Development trends

At present the energy transformation thinking brought about by climate issues and the energy supply crisis has caused North American countries to reexamine the role of nuclear energy, formulate a series of nuclear energy development strategies, and carry out many actions. Although there have been few new nuclear power units put into operation in the United States in the past 30 years, its dependence on nuclear power is still increasing. The United States has made significant progress in power plant utilization by improving the refueling, maintenance, and safety systems of existing power plants. The United States has also been placing increasing emphasis on nuclear power development, particularly during President Donald Trump's second term, when many major measures to promote nuclear energy were introduced. At the same time the United States is actively advancing next-generation nuclear power technologies. The two AP1000 reactor units in Georgia—among the first newly constructed nuclear units in the country in decades—mark an important revival of US nuclear power construction. Canada released its SMR roadmap in December 2018. Meanwhile, refurbishment projects are being carried out at the Darlington and Bruce nuclear power stations in Ontario, enabling them to continue operating beyond the mid-21st century. In October 2025 US President Donald Trump expressed a willingness to resume nuclear testing, a move widely regarded as a concerning signal regarding the potential military development of nuclear weapons.

4.3.2 Specific development of North American nuclear science and technology

4.3.2.1 Military field

4.3.2.1.1 Land-based nuclear weapons

In terms of land-based strategic nuclear forces the United States possesses 450 silo-based "Minuteman III" ICBMs, each with a range of approximately 13,000 km.

4.3.2.1.2 Sea-based nuclear weapons

Regarding sea-based nuclear forces, the United States operates 14 "Ohio-class" ballistic missile nuclear submarines, which together deploy 288 "Trident II D5" submarine-launched ballistic missiles, carrying approximately 1047 nuclear warheads.

4.3.2.1.3 Air-based nuclear weapons

In terms of air-based strategic nuclear forces the United States maintains 18 B-2 bombers and 76 B-52H bombers capable of delivering nuclear strikes. The United States is also developing a new generation bomber—the Long-Range Strike Bomber—expected to achieve initial operational capability around 2030. The plan is to procure 100 aircraft, some of which will possess nuclear strike capability.

4.3.2.1.4 Evaluation

The United States is the only country in North America possessing nuclear weapons, and it was also the first country in the world to develop them. At the end of World War II in the 1940s the United States dropped two atomic bombs on Japan, leaving a profound historical impact. Under the New START Treaty the United States agreed to reduce its deployed strategic nuclear warheads to 1550 and strategic delivery systems to 800. As of 2025 the United States is accelerating the modernization and replacement of its strategic nuclear forces to maintain the credibility and effectiveness of its nuclear deterrent. In November 2025 President Trump twice announced intentions to resume nuclear testing, casting a new shadow of global nuclear arms competition.

4.3.2.2 Energy and power field

4.3.2.2.1 Nuclear power development in the United States

The United States is the world's largest producer of nuclear power. It operates 93 nuclear reactors, accounting for about 24% of global nuclear electricity generation, and nuclear power provides more than 50% of the country's clean electricity. The US government has long supported the development of nuclear energy. After Donald Trump's return to the White House, particular emphasis has been placed on nuclear energy, with numerous policies introduced to promote its development. In 2020 the United States launched the Advanced Reactor Demonstration Program (ARDP), with an initial funding of 160 million US dollars to support the construction of two demonstration advanced reactors that can be put into operation within 5–7 years. In 2018 the Generation III nuclear power AP1000 was successfully connected to the grid for the first time. The 2021 US Infrastructure Act provides for investment in advanced nuclear reactor demonstration projects and clean power standards. In addition, the design certification of SMRs is also in progress. The United States also announced a 25 million US dollar nuclear futures package aimed at promoting modern nuclear power generation and innovative development. In 2025 President Donald Trump signed the executive order "Revitalizing the Nuclear Industrial Base," injecting strong momentum into the development of nuclear power in the United States. In the same year, nuclear companies from the United States, France, and Sweden signed an agreement to jointly explore a new pathway based on the closed nuclear fuel cycle, with a $2 billion plan focusing on liquid-metal reactors. Moreover, in 2025, First American Nuclear (FANCO) announced an ambitious nuclear energy development program aimed at advancing innovative lead-bismuth-cooled fast reactor technology.

4.3.2.2.2 Nuclear power development in Canada

Canada's CANDU energy is the only company that has successfully commercialized and scaled up heavy water reactors. CANDU reactors use natural, nonenriched uranium as fuel, but after modification, they can also use enriched uranium, mixed fuel, or even thorium. In addition, the reactor can add fuel while operating at full power, unlike most other reactors that must be stopped before adding fuel. At present, Canada has 22 CANDU reactors, 19 of which have been fully commissioned. These nuclear power units supply approximately 16% of Canada's electricity demand.

4.3.2.2.3 Evaluation

The United States and Canada in North America are both states possessing nuclear power plants, widely applying nuclear science and technology in the energy and power fields, and vigorously developing clean energy power generation. After years of relative stagnation in nuclear power the AP1000 reactor designed by the United States is actively expanding and developing.

4.3.2.3 Nuclear technology field

4.3.2.3.1 Applications in industry

Nuclear technology in the United States enjoys high economic benefits in industry applications such as semiconductor processing, meridian-irradiated tire vulcanization, and medical supplies disinfection. Sterigenics, the world's largest irradiation sterilization chain enterprise, has 40 service centers in North America, out of which 21 use irradiation methods in whole or in part. Isomedix, a chain enterprise for irradiation disinfection and sterilization, has 21 disinfection and sterilization service centers in North America.

4.3.2.3.2 Applications in agriculture

The application of nuclear technology in agriculture in the United States began early and is technologically mature. It covers key areas such as mutation breeding through irradiation, radiation-induced sterile insect techniques (SITs) for pest control, and irradiation preservation of agricultural products. Both the industrial scale and technological level of these applications rank among the world's leading positions. In addition, the impact of ocean acidification is serious; therefore the United States has established a subcommittee on ocean acidification in the Ocean Carbon and Biogeochemistry Program (OCB) secretariat to coordinate research on the issue of ocean acidification nationwide [10]. At present, radioactive isotopes are often used as tracers to detect the response of marine organisms to ocean acidification caused by carbon dioxide emissions. Using nuclear and isotopic techniques to monitor the impact of acidification and other stressors on marine ecosystems can help improve current policies, protect marine species and human life, and promote economic development.

4.3.2.3.3 Applications in the medical industry

In North America, especially in the United States, advanced nuclear technologies such as isotopes and radiation technology have been widely used in medical treatment. The US government stipulates that hospitals with more than 200 beds must have nuclear medical facilities. There are now more than 7000 hospitals that use radioactive drugs, receiving 20 million patients for diagnosis or treatment using radioactive drugs each year. In addition, according to estimates from the US Nuclear Regulatory Commission (US NRC), the United States uses radioisotopes in approximately 7 million–13 million diagnostic procedures and 550,000–650,000 therapeutic procedures each year. In addition, about 100 million laboratory tests using radioactive isotopes are conducted in the United States every year, more than 80% of which are conducted by the National Institutes of Health.

4.3.2.3.4 Applications in the environmental protection field

At present the United States has established an experimental device for treating waste gas using the flue gas desulfurization and denitrification technology by electron beam (EBA). The United States demonstration workshop has shown that this system is typically able to remove sulfur dioxide with a total removal efficiency of over 95% and a removal efficiency of nitrogen oxides reaching 80%–85% [11]. In addition, the US Defense Nuclear Agency is developing an accelerator with a power range of 0.8–1.8 MW for controlling air pollution.

4.3.2.3.5 Evaluation

Nuclear technology is fully applied in the fields of industry, agriculture, medical treatment, and environmental protection in North America, reflecting the important role of nuclear technology in North America.

4.3.3 Impact factors on development of North American nuclear science and technology

4.3.3.1 Economic development factors

North America, especially the United States and Canada, has a developed economy and abundant uranium resources, providing convenience and conditions for nuclear science and technology development. The United States is one of the world's largest producers of uranium. According to the US Geological Survey, as of 2019, the proven uranium reserves in the United States were 320,000 t. Canada is rich in uranium resources, with the world's fourth largest uranium reserves, with proven uranium reserves of 493,900 t, which accounts for about 9% of global reserves. Canada is also the second largest uranium producer in the world, with a production of 14,800 t in 2015, accounting for 15% of the total global natural uranium production. At the same time, Canada exports 12,600 t uranium, accounting for 85% of uranium production.

4.3.3.2 Environmental protection factors

Nuclear science and technology can promote environmental protection. The industrial economy in North America, especially in the United States, is well developed, and research on environmental governance and waste gas treatment is relatively in-depth. The applications of nuclear science and technology have been carried out in industrial waste gas treatment, air purification, etc., in North America.

4.3.3.3 Science and technology factors

The rapid development of science and technology in North America has made remarkable achievements in promoting nuclear science and technology, and the application of advanced nuclear science and technology can promote the development of North America. The Canadian Nuclear Safety Commission (CNSC) and the United States NRC have been actively promoting the research and development of nuclear science and technology. Canada has the technology of pressurized heavy water reactor (PHWR), which safely and efficiently converts nuclear energy into electricity. In addition, Canada has achieved efficient, safe, and sustainable development of the nuclear fuel cycle through continuous innovation and technological advancement, which has greatly promoted the development of nuclear science and technology in Canada. X-energy, an American energy company, has completed the basic design of the small high-temperature gas-cooled reactor Xe-100 and used natural uranium to manufacture the tristructural isotropic (TRISO) nuclear fuel particles. The Westinghouse Gemini robot was used for overhaul at the Beaver Valley nuclear power plant in the United States, which was a successful example of technological innovation in the nuclear industry.

4.3.3.4 Social development and public psychology factors

After the Three Mile Island nuclear accident, public sentiment has undergone significant changes, with many people switching to resist the development and utilization of nuclear energy [12]. People in major cities of the United States and residents in areas where nuclear power plants are being built have held rallies and demonstrations to demand the suspension or closure of nuclear power plants. The US government had to reexamine the plan of developing nuclear power. As a result, for more than 30 years, the United States did not build any new nuclear power plants. This also led to a new stage in people's understanding of technological risks, highlighting the crucial role of people in modern industrial technology systems. The Three Mile Island accident [13] became a classic case of risk assessment and human factors engineering and is of great significance in the history of technological thought.

4.3.3.5 National policy and governance factors

To cope with climate change the main countries in North America have more consistent goals in the development of clean and low-carbon energy, such as the US Strategic Vision report, Canada's SMRs roadmap and action plan, and so on. Although wind and solar energy are expected to lead the global energy sector to achieve zero net emissions, because of their random fluctuations, flexible, and dispatchable resources such as nuclear energy are still needed as important supplements.

4.3.3.6 Military factors

As the only country in North America that possesses nuclear weapons the US Senate voted 51–48 on 13 October 1999 to reject the Comprehensive Nuclear-Test-Ban Treaty (CNTBT) and has regarded nuclear weapons as the most effective national strategic deterrent. In November 2025 US President Donald Trump announced the resumption of nuclear testing. The United States is currently the country with the largest number of nuclear weapons in the world, as well as the most comprehensive and advanced nuclear technology, nuclear facilities, and nuclear tests.

4.4 The impact of nuclear science and technology on South America

4.4.1 General development of South American nuclear science and technology

4.4.1.1 Development status

The use of nuclear technology in South America dates back to the mid-20th century. After years of doubt and hesitation, South American countries, led by Argentina, have decided to adopt an active and prudent approach and policy to develop nuclear energy [14], aiming to independently introduce advanced nuclear energy technology from abroad, cultivate local researchers, and gradually gain experience to achieve the self-sufficient goal of nuclear science and technology. Argentina is the first South American country to use nuclear power for electricity generation and has a relatively complete nuclear fuel cycle industry. Argentina is one of the few countries in this region that are able to export nuclear equipment. Brazil is the only country in South America other than Argentina that has nuclear power plants and currently has one plant containing two pressurized water reactors. Argentina and Brazil also lead the development of nuclear technology applications in South America and have successfully applied nuclear technology to medicine, agriculture, and other fields.

4.4.1.2 Development trends

Argentina has been leading the nuclear energy development in South America and currently has three heavy water nuclear power plants in operation. Argentina is also actively seeking cooperation with China to build new nuclear power plant units. Brazil is considering the construction of new nuclear power plants and actively developing nuclear propulsion projects. Other countries in South America, such as Cuba, Chile, Ecuador, Venezuela, Bolivia, Peru, and Paraguay, have also included nuclear energy in the discussion of their policy options.

4.4.2 Specific development of South American nuclear science and technology

4.4.2.1 Military field

In the 1980s the Brazil Navy activated a nuclear propulsion program, with centrifuge development scheduled to be completed by 1989. The low-enrichment uranium demonstration plant built at the Alamar Test Center in Brazil can provide nuclear power submarines with enriched uranium of an abundance below 20%. With the help of uranium enrichment technology from the Alamar Test Centre, Brazil Atomic Energy Industry built a uranium enrichment plant in Resende. The Resende plant aims to provide enriched uranium for Angola nuclear power plants. Brazil Atomic Energy Industry Corporation fuel element manufacturing plant is also located in Resende and can produce 160 t of UO_2 pellets and 280 t of fuel assemblies per year.

4.4.2.2 Energy and power field

4.4.2.2.1 Nuclear power development in Argentina

Argentina currently has three nuclear power units in operation, including the 335 MW PHWR Atucha 1, the 692 MW Atucha 2, and the 635 MW Embalse CANDU-6. In November 2015 China National Nuclear Corporation (CNNC) and Nucleoeléctrica Argentina S.A. (NA-SA) signed two cooperation agreements in Turkey with a total value of up to 6 billion RMB. One project concerns Argentina's fourth nuclear power plant, namely, Unit 3 of the Atucha nuclear power plant, a heavy-water reactor with an installed capacity of 750 MW. The other agreement is a framework cooperation agreement for Argentina's fifth nuclear power unit.

On February 1, 2022, NA-SA officially signed the engineering, procurement, and construction (EPC) contract for the Atucha Unit 3 project with CNNC, China Zhongyuan Engineering Corporation, and China Zhongyuan Engineering Corporation Argentina Branch. CNNC will construct the Atucha Unit 3 nuclear power plant for Argentina under a turnkey EPC model.

The Atucha Unit 3 project is expected to create more than 7000 jobs, contributing to Argentina's economic and social development and benefiting the Argentine people. Argentina's selection of China's "Hualong One" reactor technology is also likely to have a demonstration effect in the highly competitive international nuclear power market, potentially influencing future international nuclear cooperation.

In addition, as a representative of indigenous innovation, Argentina's CAREM-25 project is a pioneer in the development of SMRs worldwide. It is the first integrated pressurized water reactor (iPWR) modular prototype to enter construction globally, reflecting the ambition and capability of developing countries to advance nuclear energy.

Argentina also possesses a relatively complete nuclear fuel cycle industrial chain. According to the 2020 edition of the Uranium Resources "Red Book," jointly released by the OECD Nuclear Energy Agency and the IAEA, Argentina has identified uranium resources of approximately 11,000 t. The country has set a resource exploration target of 55,000 t of uranium.

Argentina began uranium exploration and limited mining in the mid-1950s. However, because of economic considerations, uranium mining operations were halted in 1997. Up to now, Argentina has produced a total of 2582 t of uranium from seven open-pit and heap-leach mines.

4.4.2.2.2 Nuclear power development in Brazil

In 1932 scientists at the University of São Paulo in Brazil conducted nuclear fission experiments. Brazil began research on the nuclear industry in the 1950s and redefined new policies to expand the application of nuclear power in the early 1970s. It plans to build eight 1300 MW nuclear power plants by 2000. Because of the economic crisis in Latin America, including Brazil, Brazil slowed down its nuclear power development process in the early 1980s. Currently, Angra nuclear power plant is the only one in Brazil. The Central Nuclear Almirante Álvaro Alberto nuclear power plant complex in Rio de Janeiro state consists of two pressurized water reactors. Angra-1, which was first commissioned in 1985, has a power of 626 MW, whereas Angra-2, which was commissioned in 2000, has a power of 1350 MW. Brazil has three major uranium mines, namely, the Caldas mine in Minas Gerais, the Caetite mine in Bahia, and the Santa Quiteria mine in Ceará. The Brazilian government encourages the National Atomic Energy Industry Corporation to cooperate with private enterprises to increase the production capacity of the Santa Quiteria uranium mine to produce 1600 t per year by 2024. The increase in uranium mining capacity will promote nuclear power development.

4.4.2.2.3 Evaluation

In general the nuclear power development in South America is relatively slow; only Argentina and Brazil have nuclear power plants and nuclear fuel development. Chile, Cuba, and other countries have the intention to develop nuclear power.

4.4.2.3 Nuclear technology field

4.4.2.3.1 Applications in agriculture

Argentina has widely applied the characteristics of atomic radioactive isotopes to improve agricultural products, food preservation, pest control, soil trace element testing, and other aspects. In addition, the technicians of the National Atomic Energy Commission of Argentina have carried out research and a large number of applications on irradiation sterilization of dried fruits and food additives.

4.4.2.3.2 Applications in the medical industry

The Argentine Proton Therapy Center is the first nuclear medical research center in South America, which is dedicated to providing cancer treatment using proton therapy, a form of radiotherapy. Nuclear medicine in Brazil has developed for more than 40 years, and medical radioisotopes are produced at the Institute for Nuclear and Energy Research in the cyclotron center and reactor research center [15]. The spread of PET in Brazil began in the early 21st century. Although there are only two PET radioactive pharmaceutical facilities in Brazil, their production is equivalent to 6.7% [16,17]of more than 30 PET facilities in the United States.

4.4.2.3.3 Evaluation

The nuclear technology development in South America has been uneven. Because of the influence of the economic crisis and the period of the mid-1940s to early 1990s, only a few countries, such as Argentina and Brazil, have made significant use of nuclear technology in medicine and agriculture.

4.4.3 Impact factors on development of South American nuclear science and technology

4.4.3.1 Economic development factors

Economic development is one of the factors affecting the development of nuclear science and technology in South America. Because of the influence of the economic crisis and the period of the mid-1940s to early 1990s the nuclear industry in South America has developed relatively slowly. So far, only Argentina and Brazil have nuclear power plants. Compared to other energy sources the construction of nuclear power plants has a significant cost premium; therefore the high investment in nuclear power plants may exceed the resources available to most countries in South America and cannot be compensated by the expected benefits or proven to be cost-effective. In this case, such countries may decide to postpone their plan to develop nuclear science and technology.

4.4.3.2 Social development and public psychology factors

Nuclear science and technology have special requirements for the country's infrastructure to successfully introduce and safely and reliably use nuclear science and technology. These requirements are particularly important for South American countries, as most of them do not have the resources or capabilities to meet these requirements, and the psychological impact of world nuclear accidents on the people has become a major constraint on their nuclear science and technology development.

4.4.3.3 National policy and governance factors

The support of national policies and the reliability of management are the key to achieving sustainable development of nuclear science and technology. Nuclear energy is a sensitive form of energy, which is strictly controlled by nations. On the premise of ensuring sufficient nuclear safety the intention of the governments of South American countries intends to include nuclear energy in their discussion of policy options. Whether South America can achieve the sustainable development goal depends largely on the government's attitude toward the development and utilization of nuclear energy.

4.5 Impact of nuclear science and technology on Africa

4.5.1 General development of African nuclear science and technology

4.5.1.1 Development status

The severe shortage of power generation capacity and low electrification rate of residents in Africa, as well as the high electricity price and poor power supply stability, have seriously restricted the economic development of Africa. As an energy source that can generate stable basic power load, nuclear energy has great potential in alleviating energy poverty in Africa. To date, South Africa operates the only nuclear power plant in Africa, the 1940 MW Koeberg. In 2025 the African continent currently has 10 research reactors in operation, distributed in Algeria, Ghana, Morocco, Nigeria, and other countries. Africa has abundant uranium resources, with huge reserves, widespread distribution, and great mining potential, and is fully capable of supporting local nuclear power projects. Since the beginning of the 20th century, Africa has been one of the world's most important uranium mining regions. African uranium reserves account for about 20.39% of the world, mainly distributed in Niger, Namibia, South Africa, Tanzania, Botswana, and other countries [18,19]. In addition, there are also uranium resources with commercial value in Algeria, Central African Republic, Gabon, Malawi, Mali, Zambia, Zimbabwe, and other countries.

4.5.1.2 Development trends

African countries are in a very unstable situation in terms of reliable energy supply. To solve this energy crisis, some African countries have proposed to seek and develop nuclear energy capacity. Algeria, Egypt, Ghana, Kenya, Libya,

Morocco, Namibia, Nigeria, Senegal, and Tunisia have declared their intention to develop nuclear power in the near future, considering the needs of African countries for environmental protection and energy supply. The South African government announced a new round of nuclear power development plans, with the overall goal of increasing nuclear power generation capacity to 9600 MW by 2030 and gradually increasing the nuclear energy proportion in the country's total energy consumption to 30%. In addition, South Africa has decided to establish a nuclear fuel cycle system. Some African countries [20–23], such as Nigeria and Kenya, have begun the process of building nuclear power plants and training nuclear scientists.

4.5.2 Specific development of African nuclear science and technology

4.5.2.1 Energy and power field

4.5.2.1.1 Nuclear power development in Northern Africa

Egypt proposed a nuclear power development plan in 2006, initially planning to build a nuclear power plant with a power of 1000 MW in El-Dabaa, with an estimated cost of 1500 million to 2000 million US dollars. After several modifications the plan eventually evolved into the construction of four 1200 MW nuclear power plants by 2025. However, the project has not yet been completed, and the plan is to increase nuclear power generation to 9% of the country's total electricity production by 2030. In 2010 the Egyptian government passed a nuclear energy bill to provide legal support for the future large-scale nuclear power development. In 2011 the domestic situation in Egypt became increasingly turbulent under the influence of the Arab upheaval, and its nuclear power development plan was forced to be suspended. In 2013 Egypt's nuclear power plan was restarted, and the bidding for nuclear power plants was carried out. Algeria launched a nuclear energy development plan in 2007, planning to complete the first nuclear power plant construction by 2025. The plan has been postponed, with the first nuclear power plant now scheduled to be completed in 2026. The country has also established a nuclear engineering institute for it. The Moroccan government launched the country's nuclear development plan in 2007 and established the Nuclear Energy and Radiation Safety Agency in 2014. Morocco also plans to cooperate with China to build a 10 MW thermal reactor for the seawater desalination project to produce 8000 t of fresh water per day.

4.5.2.1.2 Nuclear power development in Southern Africa

In southern Africa, South Africa is currently the only country with nuclear power technology. The country's installed nuclear power capacity is 1830 MW, accounting for about 5% of the country's total power generation. South Africa is actively committed to maintaining the existing nuclear power units and extending their service life. In addition, in 2007, the president of Nigeria,

Umaru Yar' Adua, announced a nuclear energy strategy and roadmap to meet the growing demand for electricity. In 2008 Kenya's Energy Minister Kiraitu Murungi announced that the country was seeking to purchase nuclear technology to establish small nuclear power plants. In 2012 the Nuclear Electricity Board was established, with plans to start the first nuclear power plant construction by 2017 and achieve a nuclear power generation capacity of 2600 MW by 2033. Nevertheless, the nuclear power program remains unfinished, and the original timetable has been significantly delayed. Ghana established the School of Nuclear and Allied Sciences in 2006, launched a nuclear power development plan in 2007, and established the Nuclear Power Program Organization in 2012. Three nuclear power plant sites have been identified, and it is planned to increase to 1000 MW by 2025. In 2017 CGN Power Group and the government of Kenya signed a cooperation agreement on the Hualong One nuclear power plant project. In 2009 Uganda also established the Atomic Energy Commission and joined the countries developing nuclear power. The Sudanese government officially announced the launch of a nuclear energy development plan in 2007 and hoped to build a research reactor by 2020 and a 4400 MW nuclear power plant by 2030. The African Union established the African Commission on Nuclear Energy (AFCONE) in 2010 to coordinate and promote nuclear energy development plans in African countries. In 2010 Senegal also announced that it would develop nuclear power in the future. The African continent set off a wave of nuclear power development and began to sign cooperation agreements with Russia, France, the United States, and China. However, because of the weak ability for independent nuclear power development and the time from blueprint to practice for nuclear power development being relatively short, its development effect remains to be further investigated.

4.5.2.1.3 Uranium resource development

After the end of World War II, South Africa, Zambia, Gabon, Niger, and Namibia began uranium mining in the 1950s–1970s. In 1956 South Africa had become the world's second-largest uranium mining country. With the rising trend in the global uranium market, the uranium mining industry in Africa is also expanding. By 1980 the export of uranium ore in Africa had reached 17,000 t, accounting for 76% of the total world export amount [18]. After the 1980s changes in the world nuclear power market and the international political and economic situation severely restricted the African nuclear industry, which entered a low tide. The uranium exploration activities in Mali, Botswana, Namibia, Tanzania, and other countries have tended to stagnate. In addition, several newly discovered uranium mines in countries such as Namibia and Malawi are idle because of their lack of commercial value, and many uranium mines that have been put into production have been forced to stop because they cannot recover operating costs. The export of uranium mines in Africa has experienced a widespread decline in production.

As of 2015, Africa had discovered uranium reserves of about 2.81 million tons, distributed across 20 countries, accounting for approximately 21% of the world's recoverable uranium resources. By 2022 Africa had identified at least 1.068 million tons of uranium reserves, located in several countries—particularly Namibia, Niger, South Africa, Malawi, and Tanzania—representing about 18% of the world's recoverable uranium resources [19]. By 2025 with deeper exploration activities and the discovery of new deposits, Africa's annual uranium production will have exceeded 10,000 t, accounting for roughly 20% of global production. With capacity expansion in major uranium-producing countries such as Namibia, Africa's uranium output and its share of global production are expected to increase further.

4.5.2.1.4 Evaluation

In general, Africa has been slow to develop nuclear power. Across the continent, only South Africa has nuclear power plants; other countries only have research reactors, show interest in developing nuclear power, or are actively seeking cooperation with countries exporting nuclear power techniques. At present, the African uranium market has declined, but it is still an important uranium mining place in the world.

4.5.2.2 Nuclear technology field

4.5.2.2.1 Applications in agriculture

Nuclear technology is considered a way to enhance the management of agricultural production and productivity in Africa. These technologies have transformed agriculture for small- and medium-sized family farms and commercial farms. Plant mutation breeding and related biotechnology have accelerated the climate adaptability and safety of high-yield food crops and feed crop varieties. In Africa, nuclear technology is also used to ensure food safety, production, and quality control, such as conventional isotopic and analytical techniques for inspecting food traceability and authenticity, contaminant and residue analysis, and using irradiation methods to control chemical hazards in food. Nuclear technology is used to assess the availability and quality of water resources. For example, in West Africa, nuclear technology-guided drip irrigation and intelligent fertilization projects have been carried out. Nuclear technology is also applied to control various pests and diseases affecting animals and humans. A notable example is the SIT to eradicate flies and trypanosomiasis in animals and humans. The Joint FAO/IAEA Centre of Nuclear Techniques in Food and Agriculture has achieved some milestones in the applications of nuclear technology in African agriculture, such as the use of nuclear technology in Uganda to develop new cassava varieties to control cassava brown streak disease, which can both resist viruses and produce high yields.

4.5.2.2.2 Applications in the medical industry

The nuclear medicine development in Africa is relatively slow. African countries are actively seeking assistance from international organizations to develop nuclear medicine. On February 4, 2022 World Cancer Day, the Director General of the IAEA, Gloria Grossi, proposed the Rays of Hope at the headquarters of the African Union. The initiative aims to integrate the agency resources; make full use of international partnerships with member states, other international organizations, private sector, nongovernmental organizations, etc.; and assist low- and middle-income developing countries in improving cancer diagnosis and treatment capabilities through helping member states establish infrastructure, procure radiological diagnostic equipment, train professionals, and implement projects using nuclear medicine technology such as radioactive diagnosis and treatment.

4.5.2.2.3 Evaluation

The application of nuclear technology has achieved initial success in some African countries. Many African countries are actively cooperating with the IAEA on the development and applications of nuclear technology. Many countries are also assisting nuclear science and technology development in Africa, but there is still a long way to go before large-scale application.

4.5.3 Impact factors on development of African nuclear science and technology

4.5.3.1 Science and technology factors

Because of the lack of sufficient scientific and technological resources in Africa, development of nuclear science and technology has been highly dependent on cooperation with other countries. All African countries, including South Africa, have sought the help of nuclear power nations such as the United States, Russia, and France. Except for South Africa, other countries have not cultivated qualified nuclear research teams. African countries usually cannot independently carry out uranium mining. The operation of uranium mines in Niger, Namibia, Malawi, and other countries is basically dependent on foreign companies such as Australian Paladin Energy and French AREVA.

4.5.3.2 Social development and public psychology factors

Africa uranium mining activities began in the 1940s, right after which research and development of nuclear technology were also carried out simultaneously 10 years later. By the 1990s, Africa had become one of the world's most important uranium mining regions. In 1995 uranium production of Gabon, Niger, and Namibia accounted for 25% of the world's uranium mining exports. However, the downstream development of nuclear power industry in African countries is still almost in a blank state. Egypt, Libya, and Ghana have only built a few small power research reactors. After the completion of two 930 MW nuclear reactors in 1976, South Africa has not expanded the scale of nuclear power generation.

4.5.3.3 *National policy and governance factors*

African countries do not have a necessary nuclear safety management mechanism, making the risk of nuclear proliferation very serious there. In addition to the shortage of technology and professionals, African countries also have significant loopholes in nuclear safety systems. Because of the lack of corresponding institutional supervision, there have been several serious nuclear smuggling incidents in Africa. For example, in the 1970s, Libya illegally purchased 1200 t of uranium ore from Niger for its nuclear weapon research.

4.5.3.4 *Military factors*

The security environment has always been an important factor affecting the nuclear science and technology development in Africa. During the period of the mid-1940s to early 1990s, complex geopolitical environment and the resulting traditional security threats became an important incentive for countries such as South Africa, Egypt, and Libya to pursue nuclear weapons and develop nuclear technology. At the same time, various wars and turmoil seriously restricted the development of nuclear science and technology in Africa. For example, in 1982, Umkhonto we Sizwe, the armed wing of the African National Congress (ANC), attacked the Koeber nuclear power plant in South Africa, which was still under construction. It caused at least 500 million US dollars in economic damage and delayed the plant's construction schedule by 18 months.

4.6 Impact of nuclear science and technology on Oceania

4.6.1 General development of Oceanian nuclear science and technology

4.6.1.1 *Development status*

Oceania has abundant uranium resources, with Australia ranking first in the world in uranium reserves, at about 646,000 t, accounting for 41% of the world's total uranium reserves. Currently, Australia is the third largest uranium mining country in the world, with an annual production of more than 7000 t, accounting for about 10% of the global nuclear production. Although Oceania owns rich uranium resources, its nuclear power technology is very weak, with no commercial nuclear power plant built. As the largest country in Oceania, Australia passed the Australian Radiation Protection and Nuclear Safety Act in 1998, which explicitly prohibits the construction or operation of various nuclear energy facilities, including nuclear fuel factories, nuclear power plants, nuclear fuel (uranium) enrichment plants, and reprocessing plants. Although several nuclear power plant concepts have been proposed since the 1950s, Australia currently has only one nuclear research reactor located in Sydney. On the other hand, coal-fired power generation still accounts for the vast majority of the country's total power generation.

4.6.1.2 *Development trends*

Oceania has strict control over nuclear energy and mainly applies nuclear technology in industry, medical care, and other fields. But recently the topic of restarting nuclear energy has begun to emerge in Oceania. Australia has begun to actively cooperate with the United Kingdom, the United States, and other countries on nuclear energy military cooperation.

4.6.2 Specific development of Oceanian nuclear science and technology

4.6.2.1 *Military field*

The application of Oceania in the military field is mainly concentrated in Australia. In the mid-20th century the United States, the Soviet Union, the United Kingdom, France, and China successively established nuclear test bases and began nuclear tests. The United Kingdom did not have a suitable place for nuclear testing on Great Britain, so the British persuaded the Australian government to conduct nuclear tests in its region. In August 2022 Australian Defense Minister Richard Marles, accompanied by British Prime Minister Johnson, visited the nuclear submarine construction base of British Aerospace (BAE) Systems. Mars said that Australia is taking advantage of the AUKUS (trilateral security partnership among the United States, the United Kingdom, and Australia) alliance to accelerate the Australian Navy nuclear submarine program and plans to have the first nuclear submarine officially in service by 2030. In March 2023 US President Biden, UK Prime Minister Sunak, and Australian Prime Minister Albanese held the AUKUS talks in California, United States, and issued a joint statement, announcing the plan for Australia to equip nuclear submarines.

4.6.2.2 *Energy and power field*

Although Oceania does not currently use nuclear power as an energy source, Australia's Technology Investment Roadmap, released in 2020, stated that the Australian government considers SMRs a cutting-edge technology that can promote low-carbon energy transformation. In December 2019 an Australian parliamentary committee stated that the government should consider partially lifting the existing ban on nuclear energy, allowing the deployment of new technologies, including Generation III+ and Generation IV reactors. In 2021 Australia announced that it would soon cooperate with the United Kingdom in the field of nuclear technology, with SMRs identified as one of the key areas of development. In the same year, Australia and the United Kingdom signed a memorandum of understanding to cooperate in low-carbon fields such as nuclear technology and clean hydrogen energy, with future plans to develop SMRs. By 2025 this cooperation had entered a substantive phase, with companies from Australia, the United Kingdom, and the United States jointly developing offshore energy systems based on micromodular reactors, aiming to ensure that Australia has a secure and reliable energy mix.

4.6.2.3 Nuclear technology field

4.6.2.3.1 Applications in aerospace field

In Oceania, research on the application of nuclear technology in aerospace is mainly concentrated in Australia. Researchers at the Australian National University have developed a new radiation-resistant nanomaterial. The nanomaterial coating can reflect specific wavelengths of light and radiation, including far-infrared and ultraviolet radiation [24]. By calculating the vibrations of photons and applying different temperatures, researchers can enable the nanomaterial to change its refractive index, thereby reflecting different wavelengths. Compared with other materials of the same thickness, this nanomaterial provides better radiation protection, making it suitable for producing new ultrathin radiation-shielding spacesuits. Such suits could increase flexibility and mobility while improving safety for astronauts during space exploration.

4.6.2.3.2 Applications in the medical field

The application of nuclear technology in the medical field in Oceania is mainly concentrated in Australia. Since its establishment in 1969 the Australia and New Zealand Society of Nuclear Medicine (ANZSNM) has held regular meetings to provide opportunities for communication and learning among nuclear medicine professionals. The medical radiological science in Australia has undergone a long-term evolution. In 2020 the Australian Nuclear Science and Technology Organisation (ANSTO) used the Open Pool Australian Lightwater (OPAL) reactor to produce medical radioisotope scandium-47 for the first time, which can be used for the diagnosis of heart disease, kidney disease, gastrointestinal disease, and targeted cancer therapy. In 2021 Australia announced that it will build a new medical isotope molybdenum-99 production facility in Lucas Heights, where ANSTO is located, to produce molybdenum-99 by irradiating low-enriched uranium targets. At present, Australia has advanced medical equipment and professional colleges, including the Australian Society of Medical Imaging and Radiation Therapy, the Australian College of Nuclear Medicine and Radiation Therapy, and so on. In 2025 the global medical isotope supply network is shifting from a highly centralized structure toward a more diversified and regionalized system.

4.6.2.3.3 Evaluation

Radiation-protective suits are widely used in military, aviation, and other fields, and the development of protective materials has become a major research focus. In Oceania, with Australia as a representative, research on radiation-resistant materials has been actively carried out and has achieved notable results. With the continuous development of nuclear science and technology and its application in nuclear medicine, nuclear medicine has become increasingly important in the diagnosis and treatment of related diseases. Australia, as the most representative country in the development of nuclear science and technology in Oceania, has been adhering to radiation therapy technology research for many years.

4.6.3 Impact factors on development of Oceanian nuclear science and technology

4.6.3.1 Economic development factors

Countries in Oceania have been adhering to the principle of no nuclear energy for many years. The New Zealand government, the Australian Council on Climate Change (ACCC), and other organizations strongly oppose nuclear energy, mainly because of the fact that all states and territories currently prohibit nuclear energy, the construction cost of nuclear power plants is high, and the construction time is long. Wind power has an advantage in price, whereas nuclear power plants have potential safety hazards and pollution. However, small reactors have a lower investment in time and economic costs, as well as a more stable power supply. From the perspective of long-term development the development of nuclear science and technology can not only guarantee the national energy supply but also improve the competitiveness of power export and promote economic development.

4.6.3.2 Environmental protection factors

At present, nuclear power is not employed as a source of energy in Oceania. However, as the energy transformation progresses, nuclear energy has become the preferred option for countries to secure their energy supply because of its cleanliness and efficiency. Australia has announced forthcoming cooperation with the United Kingdom, with one of the focuses on the SMRs' development to ensure that Australia has a safe and reliable energy mix. When necessary, this technology can help achieve the greenhouse gas emission reduction targets set by the Paris Agreement.

4.6.3.3 Science and technology factors

There are a few studies on nuclear science and technology in Oceania, but Australia currently intends to develop emerging nuclear energy such as SMRs. However, further research and development are needed to improve nuclear science and technology capabilities and determine specific deployment routes. In 2025 the evaluation of relevant technological pathways became more specific, and the government is promoting the establishment of regulatory and investment frameworks for SMRs.

4.6.3.4 Social development and public psychology factors

Affected by the soaring international energy prices caused by the Russia–Ukraine conflict and supply chain chaos, many countries must find other reliable energy to deal with such events. There has also been a discussion on the use of nuclear energy in Oceania.

4.6.3.5 National policy and governance factors

Oceania countries are located in the South Pacific Nuclear-Free Zone. To maintain peace and security in the region, national policies oppose nuclear weapons and nuclear war. However, Australia has begun to cooperate with the United Kingdom and the United States in the research and deployment of nuclear-powered submarines recently. The New Zealand Foreign Minister said that New Zealand would not compromise on the nuclear issue and would continue to insist on the denuclearization of the South Pacific.

4.6.3.6 Military factors

The military powers in Oceania are deployed in Australia and New Zealand. Because of the surrounding ocean, they are less affected by security threats from other countries, and their military construction is mainly focused on the navy. New Zealand is actively promoting the nuclear-free zone construction in the South Pacific, whereas the Australian government is expected to purchase five US Virginia-class nuclear submarines in the future to strengthen its navy. In addition, the United Kingdom, the United States, and Australia plan to integrate their technologies and cooperate to develop a new type of nuclear-powered submarine, which is expected to be completed in 2055.

4.7 Impact of nuclear science and technology on Antarctica and Arctic

4.7.1 General development of Antarctic and Arctic nuclear science and technology

4.7.1.1 Development status

4.7.1.1.1 Antarctica

Hundreds of scientists and staff live and work in Antarctica, supporting a wide range of research, including geology, meteorology, climatology, astronomy, and so on, all of which require energy supply. As Antarctica has the lowest temperature on Earth and experiences only one sunrise and one sunset per year, it is plunged into darkness for 6 months, making most supplies undeliverable. Because of the high cost of transporting fuel to Antarctica, advanced nuclear energy may be the solution that Antarctica has been looking for. As an example the PM-3A nuclear power plant has been used to provide energy for Antarctic research stations.

4.7.1.1.2 Arctic

The current patterns of energy production and use in the Arctic rely heavily on fossil fuels. Meanwhile, another major contributor to energy production in the form of heat and electricity is nuclear energy [25]. Nuclear energy also provides power to nuclear submarines, surface warships, icebreakers, and other vessels in the Arctic of the United States, the United Kingdom, and Russia. Since 1979, people in the Arctic have also felt the impact of major nuclear accidents in the world, especially the Chernobyl accident, as well as several nuclear-powered submarine accidents in the Barents Sea and the Atlantic Ocean.

4.7.1.2 Development trends

4.7.1.2.1 Antarctica

SMRs, varying in size and scale, can generate not only energy but also heat, which is suitable for energy demands under off-grid, cold climates. These reactors can be transported by truck or airplane, making them a popular choice for remote locations. In addition to efficiency, it is important to preserve the pristine Antarctic environment, with the entire island protected by the Antarctic environment protection law, which seeks to preserve its unique ecosystem. Nuclear energy can operate in all weather conditions and is highly climate resilient. Advanced nuclear energy is sustainable, portable, and carbon-free, which is perfect for Antarctica.

4.7.1.2.2 Arctic

In the near future, Russia has special plans to expand electricity production in the Arctic through floating reactors that could provide the energy needed to expand the oil and gas industry in Russia's northern Arctic. Finland also plans to build a new nuclear power plant according to the current highest safety standards. After the Trump administration's attempt to purchase Greenland was firmly rejected, it turned instead to expanding its influence in the Arctic through strengthened military cooperation and other measures, although the acquisition plan ultimately failed.

4.7.2 Specific development of Antarctic nuclear science and technology

4.7.2.1 Scientific research field

Scientists studying Antarctic ice, which has been releasing radiation as a result of nuclear weapon tests during the period of the mid-1940s to early 1990s, have found unusually high levels of artificial chlorine in part of the ice. This chlorine stems from the nuclear weapons test that governments conducted around the world in the 1950s and 1960s, which produced high concentrations of isotopes such as chlorine-36. Radioactive isotopes from these nuclear explosions floated into the Earth's stratosphere and then began orbiting the Earth, with some of the radioactive gas eventually reaching Antarctica. In studies in the Vostok region of Antarctica, scientists have found that man-made chlorine is still present in the ice and is still being released into the atmosphere [26]. Therefore studying the chlorine content of Antarctic ice allows scientists to determine the age of the ice and better understand the history of the Earth's climate, as well as the impact of the world's nuclear weapon tests on Antarctica.

4.7.2.2 Military field

On May 3, 1958 the United States proposed to the other 11 nations participating in the International Geophysical Year (IGY) that a conference based on the consensus reached in informal discussions that the legal status quo of the Antarctic continent would remain unchanged, that scientific cooperation would continue, and the continent would be used exclusively for peaceful purposes. Negotiations culminated in the signing of the Antarctic Treaty by all 12 nations on December 1, 1959, a 12-nation agreement to protect the peaceful status of the continent, which prohibits the establishment of military bases, military exercises, deployment or testing of any type of weapon, nuclear explosions, and radioactive waste disposal in Antarctica.

The CNTBT, adopted in 1996, bans any form of nuclear explosive test in any environment on Earth and is designed to curb the development of nuclear weapons. Both the initial development of nuclear weapons and their substantial improvement (such as the emergence of thermonuclear weapons) require real nuclear tests. The CNTBT makes it hard for countries that do not yet have nuclear weapons to develop them and for those that do have nuclear weapons to develop new or more advanced ones. At the same time the treaty helps to prevent harm to human beings and the environment from nuclear tests.

4.7.2.3 Energy and power field

In the 1960s and early 1970s, McMurdo Station, the largest research base in Antarctica, was powered by a medium-sized portable nuclear reactor, PM-3A [27], the only operating nuclear power plant on the continent. Extreme temperatures of Antarctica require large amounts of fuel to maintain operations, making nuclear power an attractive alternative. The reactor's waste heat can also be used to desalinate seawater by vacuum flash distillation for the research station. Although the unit reduced the need for fuel imports, it was plagued by many problems that eventually forced its early retirement in 1972. The cost of Antarctic nuclear power made it impractical, and diesel generators have powered the base ever since. This provides a reference for the future development of Antarctic nuclear energy power.

4.7.3 Specific development of Arctic nuclear science and technology

4.7.3.1 Military field

The CNTBT bans all forms of nuclear explosive tests in any environment on Earth. Because of the rich resources in the Arctic, there is still fierce competition among some countries. Nuclear submarines of the Russian Navy are regularly cruising in the North Atlantic region. In recent years the United States has redeployed military forces in the Arctic and intensively issued various strategic documents. In 2022 the US military signed a 12-year contract with Greenland for base use to restart the "Thule Air Base" against nuclear strikes. The North Atlantic Treaty Organization (NATO) countries held the "Arctic Challenge-2023" joint military exercise in the strategic Arctic stronghold near the Russian border. In 2024 the United States redeployed military forces in the Arctic and issued a series of strategic documents in quick succession. In 2025 the US Navy officially confirmed the deployment of an E-6B Mercury aircraft to Pituffik Space Base in Greenland. The aircraft's primary mission is to maintain communications with nuclear submarines.

4.7.3.2 Energy and power field

4.7.3.2.1 Nuclear-powered icebreakers

In the 1970s the former Soviet Union began building nuclear-powered icebreakers to support activities in the Arctic [25]. The Arctic has a harsh climate, inadequate infrastructure, and difficult resource supply. Although nuclear-powered icebreakers have high construction costs and maintenance expenses, their powerful icebreaking capabilities, almost unlimited endurance, and self-sustainability in the Arctic are unmatched by any fossil-fuel-powered icebreakers. In addition, nuclear icebreakers have a lower risk of oil spills at sea than conventional icebreakers. The Russian-developed Arctic 22,220 icebreaker has a displacement of 33,500 t, and two RITM-200 reactors can provide a total power of 60 MW. By 2035 eight new nuclear-powered icebreakers are expected to join the Russian icebreaker fleet. Russia is actively developing its own fleet of nuclear-powered icebreakers with the ultimate goal of making the Northern Sea Route navigable all year round. The Northern Sea Route is a maritime shortcut connecting the Atlantic Ocean and the Pacific Ocean. It is also the third corridor in addition to the "Belt and Road" that connects the two world economic engines of East Asia and Western Europe.

4.7.3.2.2 Floating nuclear power plants

Electricity and heating have always been major problems in the Russian Arctic because of the remoteness of towns and cities, their inaccessibility, and harsh climatic conditions. On August 23, 2019 the Russian floating nuclear power plant, "Akademik Lomonosov," replaced the Bilibino nuclear power plant to provide electricity and heating. It is the world's first mobile offshore nuclear power plant, carrying two 35 MW KLT-40S nuclear reactors on a 144-m-long and 30-m-wide barge, which needs to be refueled only once every 4 years, with overhaul intervals of up to 12 years, and an endurance life of no less than 40 years.

4.7.3.2.3 Evaluation

In the field of energy power, nuclear power can greatly improve the efficiency of transportation in the Arctic, reduce the corresponding costs of insurance, labor, and fuel, and be economically competitive. Floating nuclear power plants not only solve the problem of supplying electricity and heating to remote towns and cities in the Russian Arctic, where transportation is difficult and climatic conditions are harsh. It also promotes environmental protection in the Arctic.

4.7.4 Impact factors on development of Antarctica and Arctic nuclear science and technology

4.7.4.1 Economic development factors

The Arctic and Antarctic are cold, and there are long periods of darkness. The high energy density of nuclear fuel can greatly improve the efficiency of activities in Antarctica and Arctic, especially the portable small reactors, which can reduce the corresponding transportation time, labor cost, and fuel cost and can be economically competitive while promoting the economic development of Antarctica and Arctic to a certain extent.

4.7.4.2 Environmental protection factors

Although the Arctic is relatively remote, it is not isolated from the rest of the world and is exposed to a variety of nuclear sources within the Arctic Circle and at lower latitudes. The concentrations of anthropogenic radionuclides in the Arctic are currently very low and generally declining. Nevertheless, the possible release of the K-159 nuclear-powered submarine and dumped nuclear waste may increase local contamination levels, so it is still important to conduct environmental monitoring in the Arctic. The introduction of nuclear power to the Antarctic has required and will continue to require careful assessment of the relative effects of both the plant and the environment. Environmental criteria are given, and their effect on plant design relative to radiological monitoring, secondary shielding, and radioactive waste disposal is examined for the PM-3A at McMurdo Station.

4.7.4.3 Science and technology factors

Early nuclear weapons tests in Antarctica by various countries resulted in unusually high levels of artificial chlorine in some of the ice. Although it caused environmental pollution, studying the chlorine in Antarctic ice allows scientists to determine the age of the ice and better understand the history of the Earth's climate. Advances in science and technology have motivated human beings to better study and develop Antarctic and Arctic.

4.7.4.4 Military factors

Being isolated and uninhabited the Antarctic continent could be a potential site for the deployment of nuclear weapons at some point. Over the years, there has been a general desire to keep Antarctica demilitarized, and there has been some diplomatic discussion about this possibility. Because of rich resources, there is still intense competition among certain countries across military, economic, and other strategic domains.

References

[1] B.C. Reed, A short history of nuclear physics to the mid-1930s, in: The History and Science of the Manhattan Project, Springer, Heidelberg, 2019, pp. 19–74.

[2] N. Granholm, J. Rydqvist, in: Nuclear Weapons in Europe: British and French Deterrence Forces, Swedish Defence Research Agency, Sweden, 2018, pp. 1–79.

[3] E. Rindzevičiūtė, in: Nuclear Power as Cultural Heritage in Russia, Slavic Review 80 (4) (2021) 839–862.

[4] N. Tsuzuki, Y. Kato, T. Ishiduka, in: High performance printed circuit heat exchanger, Appl. Therm. Eng. 27 (10) (2007) 1702–1707.

[5] J. Wang, et al., Design and thermal-hydraulic analysis of a printed circuit heat exchanger for ADS applications, Energy 256 (2022) 124598.

[6] W. Lei, Z. Guodong, Z. Shunping, in: Impact of Fukushima nuclear accident to China's mainland environment, Radiat. Prot. 32 (06) (2012) 325–335.

[7] CHINA ENERGY NEWS. Where Will Japan's Nuclear Power go in the Ten Years Since the Fukushima Accident [EB/OL]. [2023.07.17]. https://paper.people.com.cn/zgnyb/html/2021-03/15/content_2038429.html.

[8] C. Diww, Z. Qingbo, H. Qiong, X. Mingtao, L. Miao, Y. Qiangyi, in: Changes of cultivated land utilization pattern in Asia from 2000 to 2010, Sci. Agric. Sin. 51 (06) (2018) 1106–1120.

[9] Wikipedia, What is Mutation Breeding [EB/OL]. [2023. 07.17]. https://en.wikipedia.org/wiki/Mutation_breeding.

[10] K. Caldaria, M.E. wicket, in: Anthrop-ogenic carbon and oceanph, Nature 425 (6956) (2003) 365.

[11] M. Benjiang, in: Industrial test equipment for flue gas desulfurization and denitrification by electron beam irradiation, Environ. Prot. 2000 (8) (2000) 13–14.

[12] Science, China, Three Mile Island Accident, Baidu. 2023 [2023.7.14]. https://baike.baidu.com/item.

[13] L. Yao, Research on Three Mile Island Nuclear Accident From the Perspective of Technical Interpretation, Donghua University, Shang Hai, 2021.

[14] ATUCHA, First nuclear power station in South America, Elektrotech. Z. B 22 (12) (1970) 261–265.

[15] K.M. Bailey, Nuclear medicine in Latin America, J. Nucl. Med. 40 (9) (1999) 259–260.

[16] R. Santos-Oliveira, L.J. Antunes, Radiopharmaceutical research and production in Brazil: a 30-year history of participation in the nuclear medicine scenario, J. Nucl. Med. Technol. 39 (3) (2011) 237–239.

[17] G. Mateos, E. Suárez-Díaz, We are not a rich country to waste our resources on expensive toys': Mexico's version of atoms for peace, Hist. Technol. 31 (3) (2015) 243–258.

[18] S. Postar, The half-lives of African uranium: a historical review, Extr. Ind. Soc. 4 (2) (2017) 398–409.

[19] F. Winde, et al., Uranium from Africa—an overview on past and current mining activities: re-appraising associated risks and chances in a global context, J. Afr. Earth Sci. 129 (2017) 759–778.

[20] M. Chacha, Regional integration and nuclear energy in Africa, Afr. Secur. Rev. 21 (4) (2012) 38–50.

[21] S. Adams, S. Odonkor, Status, opportunities, and challenges of nuclear power development in Sub-Saharan Africa: the case of Ghana, Prog. Nucl. Energy 138 (2021) 103816.
[22] P.A. Trotter, M.C. McManus, R. Maconachie, Electricity planning and implementation in Sub-Saharan Africa: a systematic review, Renew. Sustain. Energy Rev. 74 (2017) 1189–1209.
[23] D. Musyoka, R.M. Field, Review of the environmental oversight framework in Kenya, in light of a nuclear power programme, Prog. Nucl. Energy 108 (2018) 89–98.
[24] M. Rahmani, et al., Reversible thermal tuning of all-dielectric metasurfaces, Adv. Funct. Mater. 27 (31) (2017) 1700580.
[25] L. Zhaoqing, G. Peiqing, From nuclear-powered icebreakers to floating nuclear power plants-Arctic nuclear energy applications in Russia, Green Technol 2020 (5) (2020) 60–63.
[26] Antarctic Ice Sheets Are Still Leaking Radioactive Chlorine From 1950s Nuclear Weapons Tests, Science alert 10 (2019). https://www.sciencealert.com/antarctic-ice-sheets-are-still-leaking-radioactive-chlorine-from-1950s-nuclear-weapons-tests.
[27] P.W. Brewer, Nuclear power in the Antarctic environment, J. Sanit. Eng. Div. 89 (4) (1963) 45–56.

Chapter 5

The human factors impact of nuclear technology

Shang Mao[a], Tao Zhou[b], Jianyu Tang[b], Chunmei Liu[b] and Wenbin Liu[b]
[a] *School of Energy and Power Engineering, Zhengzhou University of Light Industry, Zhengzhou, Henan, China,* [b] *Department of Nuclear Science and Technology, Institute of Nuclear Thermal-hydraulic Safety and Standardization, Southeast University, Nanjing, Jiangsu, China*

5.1 The impact of nuclear technology on policymakers

5.1.1 Selection of strategic development direction

5.1.1.1 The formulation and role of nuclear technology policy

5.1.1.1.1 International Atomic Energy Agency, European

The International Atomic Energy Agency (IAEA) and some European countries are promoting the peaceful use of nuclear technology worldwide, providing technical support and safety regulatory standards, and fostering cooperation and exchange among nations in the field of civil nuclear energy [1,2].

5.1.1.1.2 United States

The United States has a long history of development in the field of civil nuclear energy and operates numerous nuclear power plants. The government generally supports the application of civilian nuclear technology, viewing it as a vital source of clean energy that helps reduce dependence on fossil fuels and lower greenhouse gas emissions. For example, the US Department of Energy is committed to advancing research and development of advanced nuclear reactor technologies, such as small modular reactors, to enhance nuclear safety and economic viability. Simultaneously, it provides policy and financial support to relevant research institutions and enterprises to drive nuclear technology innovation.

5.1.1.1.3 Asia and China

Many countries recognize the positive role of nuclear technology in civilian applications, viewing nuclear energy as a key measure for achieving energy

Nuclear Science, Technology, and Civilization. DOI: https://doi.org/10.1016/B978-0-443-13941-3.00005-7
Copyright © 2026 Elsevier Inc. All rights are reserved, including those for text and data mining, AI training, and similar technologies.

diversification and addressing climate change. Several nations are actively introducing nuclear technology and constructing nuclear power plants, including Asian countries such as China, Japan, India, and South Korea. Middle Eastern nations are also promoting the peaceful use of nuclear technology globally. China has released the 3-year Action Plan for High-Quality Development of the Nuclear Technology Application Industry (2024–26), aiming to advance the application of nuclear technology in key sectors, including medical diagnostics, agricultural breeding, food processing, material modification, and security inspection. This initiative seeks to enhance independent innovation capabilities and expand industrial applications [3]. It demonstrates policymakers' emphasis on nuclear technology development and its role in promoting economic and social progress.

5.1.1.2 The role and determination of energy structure

As a clean energy source, nuclear power offers unique advantages in reducing coal consumption, lowering greenhouse gas emissions, and alleviating pressure on energy transmission. Its development is crucial for ensuring energy security and achieving the goals of carbon peaking and carbon neutrality. When policymakers consider energy structure, nuclear power—as a stable and clean energy option—can serve as a cornerstone, playing an indispensable role in realizing long-term energy security and environmental protection objectives.

5.1.1.3 Impact of decision-making in energy and environmental protection

Nuclear energy occupies a significant position in the global energy mix, particularly in addressing climate change and achieving sustainable development goals. Policymakers must consider how to support the development of nuclear energy through policy measures to optimize the energy structure and meet environmental protection objectives.

5.1.1.4 Impact of decision-making in agriculture

The application of nuclear technology in agriculture, such as food irradiation and mutation breeding, plays a vital role in enhancing food security and reducing postharvest losses. Policymakers must consider how to leverage nuclear technology to improve agricultural productivity and food safety.

5.1.1.5 Impact of decision-making in healthcare

Nuclear medicine has achieved significant advancements in diagnostic imaging and cancer treatment. Global platforms such as conferences organized by the IAEA bring together experts and policymakers to discuss progress in these fields and establish global nuclear policies and safety standards [4]. Policymakers must consider how to enhance access to radiation therapy through supportive policies to address the public health crisis of cancer treatment.

5.1.1.6 *Impact of decision-making in the military domain*

The application of nuclear technology in military domains, such as nuclear-powered submarines and aircraft carriers, significantly influences national security and military strategy. Policymakers must ensure national security while considering the appropriate application and control of nuclear technology in military contexts.

5.1.1.7 *The decision-making impact of AI computing power*

Artificial intelligence (AI) computing power finds extensive application in the nuclear energy sector. In nuclear reactor design, it enables structural optimization through simulation, significantly shortening the design cycle. Traditional methods require repeated trials, consuming considerable time. Leveraging AI computing power enables rapid parameter screening and adjustment, precise performance prediction, and enhanced design efficiency. During reactor operation, AI computing power continuously monitors massive datasets in real time, promptly detecting anomalies. Acting as an intelligent guardian, it analyzes parameters such as temperature and pressure to anticipate failures proactively, ensuring safe and stable reactor operation. Simultaneously, nuclear energy serves as a robust energy supplier for AI.

5.1.1.8 *Impact on decision-making by policymakers*

Nuclear technology exerts a comprehensive influence on policymakers, encompassing multiple dimensions such as national energy security, economic development, environmental protection, and public health. Policymakers must holistically consider these factors to formulate sound policies that foster the healthy development of nuclear technology.

5.1.2 Establishment of the legal and regulatory framework

5.1.2.1 *Definition and composition of nuclear safety laws and regulations*

5.1.2.1.1 Definition

Nuclear safety laws and regulations refer to legally binding documents such as statutes, regulations, and rules established to ensure nuclear energy safety.

5.1.2.1.2 Composition

Its composition typically includes a multitiered legal system comprising national laws, administrative regulations, departmental rules, and technical guidelines. At the international level, nuclear safety laws and regulations are primarily constituted by documents such as the Convention on Nuclear Safety and the Rules for the Notification of Nuclear Accidents and the International Nuclear Event Scale, issued by the IAEA [1]. China's nuclear safety regulatory framework comprises three tiers: laws, administrative regulations, and departmental rules. These regulations encompass the entire lifecycle of nuclear facilities—from site selection, design, and construction to operation and decommissioning—ensuring comprehensive oversight throughout the entire chain.

5.1.2.2 Scope and responsibilities of nuclear safety laws and regulations

5.1.2.2.1 General principles

The scope and responsibilities of nuclear safety laws and regulations are primarily manifested in comprehensive oversight of the entire lifecycle and entire chain of nuclear facilities, nuclear materials, and associated radioactive waste to ensure nuclear safety, prevent nuclear accidents, and minimize the potential radiological consequences of nuclear accidents.

5.1.2.2.2 US regulations

The US Nuclear Regulatory Commission (NRC) serves as the primary agency responsible for nuclear safety regulation in the United States, bearing the critical duty of establishing nuclear safety regulations, standards, and guidelines. Authorized by relevant statutes, it exercises comprehensive oversight over all aspects of civilian nuclear facilities, including the production, use, transportation, and storage of nuclear materials, to ensure public health and environmental safety. For example, the NRC's Code of Federal Regulations (10 CFR) series encompasses requirements for nuclear reactor design, operation, safety, and numerous other aspects.

5.1.2.2.3 China's regulations

5.1.2.2.3.1 Atomic energy law of the People's Republic of China
Enacted on September 12, 2025, the "Atomic Energy Law of the People's Republic of China" aims to safeguard atomic energy research, development, and peaceful utilization while advancing scientific progress and industrial upgrading. This law governs activities in relevant fields, implementing the comprehensive national security outlook and nuclear safety philosophy. The state coordinates industrial planning, implements multidepartmental oversight and supervision, provides policy support, and encourages international cooperation. It establishes provisions for scientific research and development, nuclear fuel cycles, and utilization to promote innovation, regulate industrial development, and ensure safety.

5.1.2.2.3.2 Nuclear safety law of the People's Republic of China Enacted by China on September 1, 2017, the "Nuclear Safety Law of the People's Republic of China" covers multiple aspects, including nuclear power plants, research reactors, nonreactor nuclear fuel cycle facilities, and radioactive waste management. These laws and regulations establish requirements and standards for nuclear safety, providing a legal framework for the development and operation of nuclear energy. China has established a rigorous nuclear safety policy and regulatory framework, implemented a national nuclear safety strategy, formulated medium- and long-term nuclear safety plans, and refined nuclear safety regulations and standards. These laws and regulations not only serve the development of China's domestic nuclear energy sector but also provide legal support for the country's nuclear power "going global" strategy, enhancing China's influence and voice in the international nuclear safety arena.

5.1.2.2.3.3 Law of the People's Republic of China on the Prevention and Control of Radioactive Pollution Enacted in June 2003 the "Law of the People's Republic of China on the Prevention and Control of Radioactive Pollution" is a crucial legal framework governing nuclear safety, ionizing radiation safety, and environmental protection in nuclear facilities and nuclear technology applications. It regulates the prevention and control of radioactive pollution from multiple perspectives, safeguards human health, and promotes the safe utilization of nuclear energy and nuclear technology. The 2023 revision of the Law of the People's Republic of China on the Prevention and Control of Radioactive Pollution was advanced in close alignment with industry development and regulatory practice requirements. Guided by Xi Jinping thought on the Rule of Law and China's Nuclear Safety Philosophy, this revision addresses new safety challenges arising from the expansion of nuclear energy and nuclear technology applications while integrating outcomes from ecological and environmental system reforms and practical experience from new regulations and standards in the nuclear and radiation safety sector.

5.1.2.3 Content of nuclear safety laws and regulations

5.1.2.3.1 Relevant provisions of the international nuclear nonproliferation regime

The Treaty on the Non-Proliferation of Nuclear Weapons (NPT) serves as the cornerstone of the international nuclear nonproliferation regime. Its objectives are to prevent the spread of nuclear weapons, promote nuclear disarmament, and advance the peaceful use of nuclear energy. The Treaty stipulates that nuclear-weapon states shall not transfer nuclear weapons or other nuclear explosive devices to nonnuclear-weapon states and that nonnuclear-weapon states shall not manufacture or acquire nuclear weapons or other nuclear explosive devices.

5.1.2.3.2 Accelerated Advanced Nuclear Energy Act

The US accelerated Advanced Nuclear Energy Act has taken effect, mandating the NRC to shift its mission toward balancing safety and development. It grants the NRC direct hiring authority to address talent shortages. The legislation also advances the construction of first-of-a-kind advanced reactors through specialized measures in regulation, incentives, and site selection, while reforming the nuclear technology export control system. The NRC will centralize import/export licensing; the department of energy will streamline outdated regulations; and the NRC will lead international collaboration on advanced reactor standards. In addition, related legislation prohibits Russian uranium imports to reduce US civilian nuclear sector dependence on Russia.

5.1.2.3.3 Legal provisions related to nuclear energy in China

5.1.2.3.3.1 Atomic energy law of the People's Republic of China Ensure the research, development, and peaceful utilization of atomic energy, promote scientific and technological progress, and advance industrial upgrading. Establish principles for engaging in related activities and clarify the responsibilities of relevant departments. Covering scientific research and technological development, nuclear fuel cycle, nuclear reactors and nuclear technology applications, safety supervision and management, imports and exports, and other related matters, this framework aims to drive the development of the atomic energy industry while ensuring safety.

5.1.2.3.3.2 Nuclear safety law of the People's Republic of China This legislation aims to ensure nuclear safety, prevent and respond to nuclear accidents, and promote the safe utilization of nuclear energy. It defines nuclear facilities, nuclear materials, and radioactive waste; establishes the guiding principles and policies governing nuclear undertakings; and specifies the responsibilities of entities such as nuclear facility operators. Covering multiple aspects, including nuclear facility safety, emergency response to nuclear accidents, information disclosure, and public participation, it regulates the entire lifecycle of nuclear facilities—from site selection to decommissioning.

5.1.2.3.3.3 Law of the People's Republic of China on the prevention and control of radioactive contamination Standardize the prevention and control of radioactive pollution to safeguard human health through multiple measures and promote the safe utilization of nuclear energy and nuclear technology. Establish regulations governing the prevention of radioactive pollution, emergency response procedures, and the responsibilities of polluters to ensure the safety of radioactive materials throughout production, use, and disposal processes.

5.1.2.3.3.4 Environmental Radiation Protection Regulations for Nuclear Power Plants The newly revised and released Environmental Radiation Protection Regulations for Nuclear Power Plants (GB6249-2025) officially took effect on May 1, 2025, replacing the 2011 edition. This revision aims to align with China's nuclear technology advancements and environmental regulatory requirements, further strengthening environmental radiation protection standards for nuclear power plants. This revision signifies a further deepening of China's alignment with international standards for environmental radiation protection at nuclear power plants. It reflects a high priority on public health and ecological environment while providing institutional safeguards for the promotion of new technologies such as small modular reactors. Enterprises must focus on key areas, including site selection justification, discharge monitoring, and emergency response system development to ensure compliance with the mandatory requirements of the new regulations.

5.1.2.4 Modernization of nuclear safety laws and regulations

Policymakers are committed to modernizing the regulatory system and regulatory capabilities, including systematizing, scientifically grounding, legalizing, informatizing, and refining regulatory work. This requires policymakers to continuously update and optimize nuclear safety regulations to meet the demands of technological advancement and socioeconomic development.

5.1.2.5 International cooperation on nuclear safety laws and regulations

The Nuclear Technology Application and Technical Cooperation Program brings together experts, policymakers, and stakeholders to discuss the socioeconomic benefits of nuclear technology and develop global nuclear laws, regulations, and safety standards.

5.1.2.6 Evaluation

Nuclear technology exerts a profound influence on policymakers, impacting multiple dimensions, including the formulation and enforcement of nuclear safety regulations as well as international cooperation. This necessitates that policymakers ensure the rational utilization of nuclear technology while safeguarding safety, thereby promoting sustainable socioeconomic development.

5.1.3 Policy formulation

5.1.3.1 Civil nuclear policy

5.1.3.1.1 International

The IAEA plays a vital role in promoting global nuclear safety, safeguards and verification, and technical cooperation. Through collaboration with the IAEA, countries enhance their national nuclear regulatory capabilities, ensure the safe operation of nuclear facilities, and engage in international cooperation in areas such as nuclear technology research and development and nuclear waste management.

5.1.3.1.2 China

China's civil nuclear policy adheres to the principle of "safety first, quality first," clearly defining a three-step nuclear energy strategy—"thermal reactors, fast reactors, and fusion reactors"—along with a closed nuclear fuel cycle technology roadmap. Through the formulation of 5-year plans and medium-to-long-term plans, China has established a rigorous policy and regulatory framework and implemented full-chain safety licensing and stringent technical reviews, ensuring the safety of nuclear facilities throughout their entire lifecycle. Concurrently, China actively promotes the diversified application of nuclear power, strengthens nuclear safety culture development and international cooperation, and safeguards the safe and orderly advancement of nuclear energy.

5.1.3.2 Military nuclear policy

5.1.3.2.1 National defense development and overall layout

Countries generally formulate long-term national security strategies that clearly define the objectives, tasks, and priorities of national defense development. For example, the United States' National Security Strategy Report and China's White Paper China's National Defense in the New Era provide macrolevel guidance for national defense development. China's nuclear science and technology innovation capabilities are also continuously advancing, effectively serving national defense development and the peaceful development and utilization of nuclear energy. This underscores the critical role of nuclear science and technology in national defense development.

5.1.3.2.2 The development and restriction of nuclear weapons

5.1.3.2.2.1 Restrictions of the Comprehensive Nuclear-Test-Ban Treaty

On September 10, 1996 the United Nations General Assembly adopted the Comprehensive Nuclear-Test-Ban Treaty by vote. The treaty's purpose and objective are to comprehensively prohibit nuclear weapon test explosions and any other nuclear explosions, thereby effectively promoting the comprehensive prevention of nuclear weapons proliferation and the process of nuclear disarmament. Among the 44 nuclear-capable states required for the treaty's entry into force, India, Pakistan, and North Korea have yet to sign. The United States, China, Egypt, Iran, Israel, and Russia have signed but not formally ratified the treaty.

5.1.3.2.2.2 The Chinese government's advocacy for a comprehensive ban The Chinese government maintains a consistent and resolute stance on nuclear weapons, advocating for their comprehensive prohibition and complete elimination. It has taken the lead among nuclear-armed states in making a solemn commitment to "no first use of nuclear weapons." This policy helps reduce the risk of nuclear weapon use and promotes global strategic stability. The International Nuclear Initiative represents a series of proposals and action plans advanced by the international community to address security threats posed by nuclear weapons and to advance nuclear disarmament and nonproliferation. For instance, China has long advocated for nuclear-weapon states to negotiate a "No First Use Treaty" or issue relevant political declarations. This initiative aims to reduce the risk of nuclear war and promote global strategic balance and stability.

5.1.3.2.2.3 Restrictions of the Treaty on the Non-Proliferation of Nuclear Weapons The NPT is dedicated to preventing the spread of nuclear weapons, promoting nuclear disarmament, and advancing the peaceful use of nuclear energy. The Treaty on the Prohibition of Nuclear Weapons, meanwhile, seeks to prohibit the development, testing, production, and related activities of nuclear weapons at the level of international law.

5.1.3.2.3 A new phase in great power rivalry

5.1.3.2.3.1 US Withdrawal from the INF Treaty As nuclear powers the nuclear arms control treaties between the United States and Russia are crucial to global nuclear stability. In recent years, competition in the nuclear domain has shown a trend of gradual intensification, significantly escalating strategic rivalry between major powers. The United States views the INF Treaty as a constraint on its military capabilities, particularly in the key strategic regions of Europe and East Asia. The treaty prevents the United States from deploying land-based intermediate-range missiles in these areas, undermining its strategic advantage over Russia and China. Withdrawing from the treaty will allow the United States to freely deploy intermediate-range missiles, thereby consolidating and expanding its strategic advantage.

5.1.3.2.3.2 Russia suspends participation in the New START Treaty Amid heightened tensions between the United States and Russia in recent years the New START Treaty has maintained a degree of balance in the nuclear forces of both nations. However, considering its own circumstances, Russia announced in February 2023 that it would suspend its participation in the treaty. This decision has dealt a significant setback to the nuclear arms control process and heightened the uncertainty surrounding the nuclear standoff between the two powers.

5.1.3.2.3.3 The strategic landscape is undergoing a complex adjustment process The dynamic evolution of the situation reflects profound shifts in the global strategic power structure. Against the backdrop of heightened strategic autonomy and enhanced policy effectiveness among nations the international nuclear security order and the logic of great power competition are navigating a complex realignment. This strategic reassessment by all parties extends beyond military security dimensions to encompass geopolitics, economic interests, and international rule-making—serving as a crucial window into the unfolding evolution of the current international strategic landscape.

5.1.3.2.4 US policy

The strategic adjustment in US nuclear policy is primarily reflected in the following shifts: following the Cold War, the strategy transitioned from relying solely on offensive nuclear forces for deterrence to a "dual-capability" approach, integrating strategic defense capabilities into the nuclear force structure. During the Trump administration the scope for nuclear weapons use was expanded, introducing the concept of "tailored deterrence" and relaxing restrictions on the development of low-yield nuclear weapons. In addition, in November 2025, Trump twice announced plans to resume nuclear testing.

5.1.3.2.5 Russian policy

Russia has significantly broadened the conditions under which it may employ nuclear weapons, now including severe conventional attacks against its homeland or ally Belarus as scenarios that could trigger a nuclear counterstrike. This shift marks a transition in its nuclear deterrence strategy from the previous doctrine of using nuclear weapons only when "national survival is threatened" to a more proactive, lower threshold "dose-based" deterrence model. Meanwhile, in response to US President Trump's announcement of plans to resume nuclear testing, Russia declared that to defend its national security and strategic balance, it would likely have no choice but to follow suit and restart its own nuclear testing program.

5.1.3.2.6 China's policy

China's strategic adjustment in nuclear policy, which calls for "accelerating the development of strategic deterrence capabilities," demonstrates the flexibility of China's nuclear policy and its responsiveness to changes in the international security landscape. This adjustment helps enhance China's strategic autonomy and amplify its voice in international affairs. However, China has consistently adhered to a no-first-use policy for nuclear weapons. This stance, established and maintained because China acquired nuclear capabilities, reflects China's high degree of restraint and responsible attitude toward nuclear policy, as well as its determination to uphold global strategic stability.

5.1.3.3 International cooperation and global nuclear policy-making

International cooperation serves as the cornerstone of global nuclear policy-making. Through platforms such as the IAEA, nations jointly negotiate rules and treaties with the core objectives of preventing nuclear weapons proliferation, advancing nuclear disarmament, and ensuring the safe and reliable peaceful use of nuclear energy and technology. This cooperation is built upon mutual verification and trust, aiming to balance national security with shared global interests and collectively uphold a safer nuclear order.

5.1.3.4 Cultivating nuclear safety culture

Policymakers emphasize the cultivation of nuclear safety culture, embedding its core values into the mindset and actions of personnel through ideological education, institutional norms, and environmental immersion.

5.1.3.5 Research and development of nuclear safety-related science and technology

Policymakers encourage and support the research, development, and utilization of nuclear safety-related science and technology, strengthen intellectual property protection, and prioritize the cultivation of nuclear safety professionals. This contributes to enhancing the safety and efficiency of nuclear technology applications.

5.1.3.6 The role of science and technology diplomacy

Nuclear technology plays a vital role in science and technology diplomacy, particularly in areas such as fundamental science, energy, environment, and climate change; global public health governance; intellectual property; technical standards and science and technology policy; and civil nuclear technology and nuclear safety.

5.1.3.7 Evaluation

Nuclear technology exerts a comprehensive influence on policymakers across cultural, scientific, diplomatic, and defense domains, impacting multiple dimensions, including strategic adjustments to nuclear policy, defense development, international cooperation, and the advancement and limitation of nuclear weapons. Particularly significant are relevant international nuclear initiatives, which hold vital importance for safeguarding global peace and security and building a nuclear-free world.

5.2 The impact of nuclear technology on practitioners

5.2.1 Nuclear safety regulatory activities

5.2.1.1 Purpose and value

5.2.1.1.1 Objectives of nuclear safety regulations

Nuclear safety concerns the protection of people's lives, property, and social stability. During the development of nuclear energy, accidents or leaks can cause severe impacts on human health, the ecological environment, and the socio-economy. The formulation and enforcement of nuclear safety laws, regulations, and oversight measures can reduce nuclear safety risks at their source, ensuring the stable and secure operation of nuclear energy.

5.2.1.1.2 Objectives of nuclear safety law oversight

Strengthen oversight of nuclear power operators to promptly identify and address potential safety hazards, ensuring the normal operation of nuclear facilities. Simultaneously, enhance emergency preparedness and response capabilities for nuclear incidents to improve crisis management, thereby minimizing the occurrence of nuclear accidents and their societal impact. Nuclear safety laws, regulations, and oversight measures can propel the sustainable development of nuclear energy. By establishing nuclear safety legislation, strengthening oversight and constraints on nuclear enterprises, and standardizing the development of the nuclear industry, these measures advance the sustainable growth of nuclear energy. Concurrently, enhancing nuclear technology R&D and innovation improves nuclear safety and efficiency, continuously driving the progress of the nuclear energy sector.

5.2.1.1.3 The value of nuclear safety regulation

Nuclear safety laws, regulations, and regulatory measures serve as vital instruments for safeguarding and advancing nuclear energy security. Only through the formulation and implementation of relevant regulations and measures, enhanced oversight of nuclear facilities, and improved safety and stability of nuclear technology can the secure development of nuclear energy be ensured, thereby maximizing the protection of public life and property. Consequently, high priority should be given to the development and enforcement of nuclear safety laws, regulations, and regulatory measures. Efforts to strengthen nuclear safety must be intensified to contribute to the sustainable development of nuclear energy.

5.2.1.2 Basis and process

5.2.1.2.1 IAEA

The IAEA has established a series of nuclear safety standards and guidelines covering nuclear facility design, operation, and decommissioning; nuclear material management; radiation protection; and other areas, providing reference frameworks for national nuclear technology regulation [1]. Examples include the guidelines on "safety design of nuclear power plants" and "operational safety of nuclear power plants." National regulatory systems for nuclear technology, centered on "safety, compliance, and nonproliferation," form a multitiered, multidimensional control network [13].

5.2.1.2.2 United States

The United States has established a comprehensive regulatory framework through the NRC covering the entire lifecycle of reactors—from design and operation to decommissioning. It enforces a strict licensing system for the transportation of nuclear materials and the use of highly enriched uranium. In November 2025 the Advisory Committee on Reactor Safety (ACRS) of the US NRC conducted its triennial review and assessment of the NRC's safety research program [5]. This review encompassed 11 research projects within the NRC's Office of Nuclear Regulatory Research, focusing on whether the research program meets current and future regulatory needs of the NRC. It also evaluated the program's role and impact in enabling the NRC to fulfill its mission.

5.2.1.2.3 Russia

Russia has integrated nuclear regulation under the umbrella of the State Atomic Energy Corporation (ROSATOM), emphasizing the separate management of military nuclear technology and civilian nuclear power. Concurrently, it has enacted the Nuclear Safety Act to strengthen monitoring of nuclear material smuggling across its borders.

5.2.1.2.4 European Union

The European Union implemented the Nuclear Safety Directive, requiring member states to conduct regular stress tests on nuclear power plants, establish a pan-European nuclear emergency response mechanism, and implement nuclear material accounting and control through the European Atomic Energy Community (EURATOM).

5.2.1.2.5 Japan

Following the Fukushima accident, Japan revised its nuclear reactor regulation act, introducing stricter seismic and tsunami protection standards. It established an independent nuclear regulatory authority (NRA), severing the industry's regulatory conflicts of interest.

5.2.1.2.6 China

For nuclear facilities the National Nuclear Safety Administration implements independent nuclear safety oversight across all stages—site selection, design, construction, operation, and decommissioning—encompassing technical reviews, administrative licensing, and supervisory inspections, as illustrated in Fig. 5.1.

5.2.1.3 Approach and methodology

5.2.1.3.1 Objective approach

The IAEA has established a series of key principles and methodologies in the field of nuclear safety. Its core objective is to ensure the safe and reliable use of nuclear energy while protecting human health and the environment from radiological hazards. "Safety First" serves as the central tenet of the IAEA's nuclear

FIGURE 5.1 Nuclear technology regulatory approach.

safety principles, permeating the entire process of nuclear energy development, utilization, and regulation.

5.2.1.3.2 Methodological characteristics

China's nuclear safety regulation adopts approaches consistent with international practices, characterized primarily by the following features: First, based on safety standards issued by the IAEA, China has established a systematic and comprehensive framework of nuclear safety regulations, guidelines, and technical documents, maintaining alignment with advanced international nuclear safety requirements [1]. Second the National Nuclear Safety Administration implements a phased administrative licensing system for nuclear facilities. Licensing reviews are conducted for each phase—site selection, construction, fuel loading, operation, and decommissioning—and operators may proceed to the next phase only after obtaining the relevant licenses. Third the National Nuclear Safety Administration oversees the entire lifecycle of nuclear facilities, from design, manufacturing, and construction through commissioning, operation, and decommissioning.

5.2.1.4 System and requirements

5.2.1.4.1 Regulatory framework

To enhance oversight of nuclear power plant operations a multitiered regulatory system must be established to achieve effective supervision at multiple levels. For example, China has developed a nuclear safety regulatory framework that aligns with international standards while addressing domestic conditions, covering the entire lifecycle from site selection, design, construction, and operation to decommissioning. China has established six regional nuclear and radiation safety supervision stations nationwide, dispatching on-site supervisors to nuclear power plants to conduct comprehensive oversight of nuclear facility operations. A three-tiered nuclear safety regulatory framework has been formed, comprising headquarters agencies, regional supervision stations, and technical support units. Enterprises themselves have also strengthened their nuclear safety supervision systems by refining internal oversight mechanisms and detailing prevention and control procedures to fortify nuclear safety defenses. China's nuclear power sector will continue deepening internal oversight system reforms in 2025. Seizing the opportunity presented by the "Year of Pursuing Excellence," it will establish a multitiered, full-chain oversight and management framework.

5.2.1.4.2 Specific requirements

Headquarters formulates policies and overall regulatory requirements; regional supervision stations conduct on-site inspections; technical support units provide specialized technical verification and assessment; implementers undergo unscheduled inspections by regional supervision stations during routine operations. Everything from equipment operational status and personnel procedures to safety management system implementation undergoes detailed scrutiny. Technical support units also conduct reviews of critical technical aspects, such as technical assessments of nuclear safety license applications and technical evaluations of nuclear equipment design and manufacturing. Operators' work remains under constant oversight, compelling them to maintain a high sense of responsibility and rigorous work ethic. Any negligence may be uncovered during regulatory inspections, potentially leading to severe consequences.

5.2.1.5 Evaluation

Nuclear safety laws, regulations, and regulatory measures form the core of each country's nuclear safety management system, designed to ensure the safety of nuclear facilities, materials, and activities. Regulatory authorities employ enforcement, penalties, and oversight to ensure nuclear facility operations comply with regulatory requirements. They impose warnings, fines, and suspension of operations for violations to uphold nuclear safety. Simultaneously, countries are continuously strengthening their nuclear safety regulatory capabilities, refining regulatory and standards frameworks, and enhancing regulatory technical expertise to address new challenges in nuclear energy development.

5.2.2 Site selection for nuclear power plants

5.2.2.1 Principles for selecting nuclear power plant sites

5.2.2.1.1 Principles and content

The siting of nuclear power plants adheres to four fundamental principles: technical and economic viability, safety performance, environmental safety, and social stability. It requires thorough and repeated evaluation.

5.2.2.1.2 Detailed explanation

5.2.2.1.2.1 Technical and economic considerations The site must meet the technical operational requirements of a nuclear power plant, such as proximity to water sources to ensure cooling water supply and geological stability to prevent seismic or other geological hazards from compromising facility safety. Simultaneously, economic factors must be considered: the location should facilitate power transmission, be near load centers or grid hubs to reduce transmission costs, minimize land development and relocation expenses, and ensure both technical feasibility and reasonable economic benefits throughout the project's entire lifecycle.

5.2.2.1.2.2 Safety performance The core objective is to prevent radioactive material leakage. Site selection must avoid high-risk geological zones such as active seismic belts, fault lines, and volcanoes, while steering clear of areas prone to extreme weather events such as hurricanes and tsunamis. In addition, a reasonable safety buffer zone must be maintained from densely populated areas. In the event of an accident, radiation risks to the public must be minimized, and comprehensive emergency response capabilities must be ensured.

5.2.2.1.2.3 Environmental safety Assess the impact of site selection on surrounding ecosystems to avoid damaging sensitive areas such as nature reserves, habitats of rare flora and fauna, and water source protection zones. Ensure that pollutants generated during nuclear power plant operations—including warm discharge water, cooled discharge water, and low-level radioactive wastewater—meet environmental standards after treatment. This prevents harm to water bodies, soil, atmosphere, and biodiversity while maintaining regional ecological balance.

5.2.2.1.2.4 Social stability Public acceptance must be fully considered by soliciting local residents' opinions through public notices, hearings, and other means to minimize social disputes arising from site selection. Simultaneously, site selection should avoid impacting local cultural heritage sites and critical public facilities such as airports and military installations. The normal production and living order of surrounding residents must be safeguarded to ensure social stability throughout the project's construction and operation phases.

5.2.2.2 Influencing factors

5.2.2.2.1 Safety-related factors

5.2.2.2.1.1 Environmental factors Includes characteristics that may affect site suitability, considering features that could influence the potential radiological consequences of a nuclear power plant on the environment, such as population distribution around the site and meteorological and hydrological conditions.

5.2.2.2.1.2 Extreme external factors Thoroughly demonstrate the existence of extreme external events in the plant site area that could have a severe impact on the nuclear power plant, such as geological conditions, earthquakes, tsunamis, floods, extreme weather, aircraft crashes, and chemical explosions.

5.2.2.2.2 Nonsafety factors

5.2.2.2.2.1 Socioeconomic factors These primarily include the power grid, operating conditions, topography, water supply, and other socioeconomic factors.

5.2.2.2.2.2 Phased factors China imposes stringent requirements for nuclear power plant site selection. The process generally comprises three phases: site investigation, site evaluation, and preoperation.

5.2.2.2.2.3 Site investigation The objective of site investigation is to identify one or several priority candidate sites and conduct systematic screening and comparison of these locations.

5.2.2.2.2.4 Site evaluation Site evaluation involves investigating and assessing one or more priority candidate sites to demonstrate their acceptability from a safety perspective. Concurrently, preliminary determination of site-related design bases is required.

5.2.2.2.2.5 Preoperational phase The preoperational phase involves completing and refining the evaluation of site characteristics while validating and verifying results from preceding phases. A suitable nuclear power plant site requires rigorous selection, multiple rounds of screening, and strict oversight before final confirmation.

5.2.2.3 Design methodology

5.2.2.3.1 Method types

Nuclear power plant site selection is a systematic engineering endeavor concerning safety and efficiency, requiring the integrated application of deterministic methods and probabilistic safety assessment (PSA) analysis.

5.2.2.3.2 Deterministic methods

Deterministic methods start from site selection criteria, using historical data and engineering standards to establish screening thresholds for natural conditions (e.g., earthquakes, floods, and meteorological hazards) and external human factors (e.g., aircraft crashes and chemical plant explosions) in the site area, assessing their impact on nuclear plant safety. Through long-term monitoring and analysis of geological structures, hydrological conditions, and meteorological parameters, it ensures the site can withstand extreme events without compromising nuclear safety. Simultaneously, it considers social conditions such as population distribution, transportation networks, and emergency response capabilities to guarantee effective emergency implementation during incidents.

5.2.2.3.3 Probabilistic safety assessment method

PSA adopts a risk quantification approach, employing models such as event trees and fault trees to calculate the probability of potential accident sequences at nuclear power plants. It evaluates the likelihood of core damage and radioactive releases triggered by different initiating events, along with the severity of their consequences. By integrating surrounding population density and environmental characteristics, PSA quantifies the potential risks to public health and the environment from nuclear accidents, providing probabilistic risk metrics for site selection decisions.

5.2.2.3.4 Methodological integration

Deterministic methods provide "hard constraints" to ensure sites meet minimum safety standards, whereas PSA offers "soft evaluations" that supplement deterministic approaches with risk quantification. These complementary methodologies work in tandem: the former defines safety boundaries, whereas the latter optimizes site selection. Together, they ensure scientifically sound nuclear power plant siting that balances safety and economic viability.

5.2.2.4 Evaluation

The development and utilization of nuclear energy have brought significant economic benefits to humanity, but they also pose potential dangers to the natural environment and people's social lives. Particularly after the Chernobyl nuclear accident in the former Soviet Union and the Three Mile Island accident in the United States, concerns about nuclear power safety—including nuclear power plant siting—have intensified. The nuclear leakage incident at Japan's Fukushima power plant, triggered by earthquakes and tsunamis, inevitably demands serious consideration of nuclear safety under extreme external events such as seismic activity. In reality, however, international standards for nuclear power plant siting are extremely stringent, supported by a comprehensive system designed to ensure the safety of nuclear power plant locations.

5.2.3 Technology upgrades and optimization

5.2.3.1 Material enhancement

5.2.3.1.1 Development of new materials

Utilizing advanced material technologies to develop novel materials with high strength and radiation resistance, such as high-temperature alloys and reinforced ceramic materials.

5.2.3.1.2 Enhancing material preparation processes

Improve the control over crystal structure and internal defects by refining material preparation techniques, thereby boosting mechanical properties and radiation resistance.

5.2.3.1.3 Incorporating composite materials

Replace conventional materials with composites to leverage their superior characteristics—including high strength, lightweight properties, and corrosion resistance—thereby enhancing the performance and service life of nuclear power equipment.

5.2.3.2 Manufacturing process improvement

5.2.3.2.1 Approach

The adoption of advanced technologies such as new-type productive forces is crucial for enhancing production efficiency and product quality through process improvements in nuclear power equipment manufacturing [6,7].

5.2.3.2.2 Example

Implementing digital manufacturing technologies enables the digitization, automation, and intelligent optimization of production workflows, thereby boosting manufacturing efficiency and precision.

5.2.3.3 Design optimization

5.2.3.3.1 Optimize structural design

Enhance the structural design of nuclear power equipment to improve overall stability and seismic resistance [3,7], thereby reducing accident risks.

5.2.3.3.2 Improve fuel combustion efficiency

Refine fuel rod design to increase fuel utilization, minimize nuclear waste generation, and boost the economic efficiency of nuclear power plants.

5.2.3.3.3 Reduce operational risks

Integrate advanced automation control technologies into nuclear power equipment to enhance operational stability and controllability, thereby lowering the risk of human operational errors.

5.3 The effects of nuclear technology on workers

5.3.1 Professional competency and training for personnel directly involved in the nuclear industry

5.3.1.1 Nuclear facility operators

5.3.1.1.1 Training objectives

In civil nuclear facilities, radiation exposure occurs only when personnel enter operational areas. Therefore much radiation exposure can be prevented through training for nuclear power plant and contractor personnel. This principle must never be relaxed until every worker clearly understands the core principles and concepts of "ALARA Awareness." By knowing where to stand, where to rest, which areas to avoid, and the optimal walking routes, many exposures can be easily prevented.

5.3.1.1.2 Training content

5.3.1.1.2.1 Implementing the fundamental ALARA concept and rationale Specifically, when work involves trade-offs between dose and cost, the fundamental ALARA concept must be implemented. Key rationale should be communicated to enhance workers' awareness of reducing exposure doses, improve their personal safety levels, and balance the plant's interests in cost reduction with plant safety.

5.3.1.1.2.2 Training in radiation concepts and practices Personnel must receive training in fundamental radiation protection concepts and best practices to safeguard themselves and others, minimizing exposure during both operational and nonoperational periods.

5.3.1.1.2.3 Understanding plant management methods and recommendations Beyond these core principles, personnel should be familiarized with plant management methodologies and recommendations to successfully apply them in their work and demonstrate effective performance in task management.

5.3.1.1.2.4 Learning and adopting good personal practices at work Staff should learn and adopt good personal practices at work to reduce doses and complete tasks with high quality.

5.3.1.1.2.5 Cultivating measures to take during unforeseen events Training and education should also include measures to take if unforeseen events occur during work, guiding staff on how to act in such situations.

5.3.1.1.3 Training methods

5.3.1.1.3.1 Explain the fundamental concepts of ALARA and best practices in radiation protection Regarding staff participation in ALARA practices, training involves explaining the fundamental concepts of ALARA and best practices in radiation protection, providing staff with the knowledge necessary to fulfill their duties and maintain radiation exposure at ALARA levels.

5.3.1.1.3.2 Must be tailored to the type of work and responsibilities of the personnel being trained Training must be tailored to the type of work and responsibilities of the personnel being trained. For example, specialized training sessions for managers may cover the importance and justification of implementing an ALARA framework, its fundamental principles, the necessity of establishing objectives, the importance of multidisciplinary teams participating in overhaul preparation and implementation activities, and the procedures and methods for evaluating the effectiveness of the framework's implementation.

5.3.1.1.3.3 Must cover radiation protection management techniques, specialized functions, and responsibilities for frontline personnel Training for frontline radiation protection specialists must extend beyond basic ALARA principles to include radiation protection management techniques (pre- and postwork reviews and dose reduction methods), staff responsibilities, and the specialized role of frontline radiation protection specialists as assistants and advisors to other departments and personnel. Maintenance and operational staff must understand radiation protection techniques relevant to their work.

5.3.1.1.3.4 Example The impact of working conditions on exposure duration. When establishing new work procedures, developing tools for use in controlled areas, or wearing bulky protective clothing, these factors should be taken into account.

5.3.1.2 Radiation protection personnel in nuclear medicine

5.3.1.2.1 Scope of radiation protection in nuclear medicine

Radiation protection in nuclear medicine primarily encompasses radiation protection requirements for personnel and workplaces involved in nuclear medicine diagnosis, treatment, research, and radiopharmaceutical preparation within medical institutions, as well as protection during the use of radioactive substances.

5.3.1.2.2 Radiation protection responsibilities of practicing physicians

Physicians shall ensure patients or examinees are shielded from unnecessary exposure. Their primary responsibilities and obligations include the following:

Conducting justification assessments

Informing patients or examinees of radiation health risks

Providing relevant information to other practicing physicians

Ensuring all novel nuclear medicine diagnostic/therapeutic techniques and methods undergo justification assessment before institutional adoption

Reviewing prior patient/examinee examination records before requesting radiopharmaceutical procedures

Avoiding unnecessary examinations

5.3.1.2.3 Radiation protection responsibilities of nuclear medicine personnel

Nuclear medicine practitioners, radiation protection officers, and other relevant personnel bear corresponding responsibilities for applying radiation protection regulations and standards within their specific areas of activity. Nuclear medicine physicians should understand the characteristics of relevant medical imaging diagnostic and therapeutic techniques and strictly control their scope of use. To prevent unintended radiation exposure to embryos, fetuses, and infants, patients or examinees must be questioned and assessed regarding pregnancy or lactation status. Corresponding records must be maintained, and relevant informed consent notices must be prominently displayed at the entrance to the nuclear medicine department and in the preadministration waiting area. When reviewing applications for radiopharmaceutical diagnosis and therapy, nuclear medicine physicians must implement measures necessary to adhere to the ALARA principle.

5.3.1.3 Protective measures

5.3.1.3.1 Protective measures for special populations undergoing radiopharmaceutical therapy

5.3.1.3.1.1 Protective measures during diagnosis Except when clinically indicated and the use of radiopharmaceutical diagnostic techniques is essential, diagnostic radiopharmaceuticals should be avoided for pregnant women whenever possible. If use is unavoidable, the patient or examinee should be informed of the potential risks to the fetus. Except when clinically indicated and absolutely necessary, avoid administering radiopharmaceuticals to lactating women. If administration is unavoidable, advise the patient or examinee to temporarily discontinue breastfeeding. Except when clinically indicated and absolutely necessary, radionuclide imaging should generally not be performed on children. If such an examination is required, the administered dose of radiopharmaceutical should be minimized, and radionuclides with short half-lives should be preferred.

5.3.1.3.1.2 Protective measures during treatment Unless life-saving, radioactive drug therapy should not be administered to pregnant women, particularly those involving 131I and 32P. When life-saving radioactive drug therapy is necessary, fetal dose exposure should be assessed according to special protocols, and patients should be informed in writing of potential risks to the fetus. Unless absolutely necessary to save life, radiopharmaceutical therapy should be avoided in lactating women. If administration is unavoidable, patients should be advised to temporarily discontinue breastfeeding.

5.3.1.3.2 Measures taken by nuclear medicine physicians when reviewing applications for radiopharmaceutical diagnosis and treatment

1. Select appropriate radiopharmaceuticals and their administered activity based on the patient's or examinee's physical condition, paying particular attention to children and patients/examinees with impaired organ function.
2. Employ methods to block radioactive drug uptake in nontarget organs and accelerate their elimination.
3. Fully utilize existing information to avoid all unnecessary repeated exposures, minimizing the dose received by patients or examinees.
4. Refer to corresponding nuclear medicine diagnostic reference levels. Establish procedures to verify radiopharmaceutical treatment plans and the identity of patients or examinees.
5. Upon discharge, provide written and verbal guidance to patients meeting radiopharmaceutical requirements to effectively minimize exposure to family members, caregivers, and the public—particularly minors and pregnant women.

5.3.1.4 Management of medical radioactive waste

5.3.1.4.1 Scientific classification and management of radioactive waste

5.3.1.4.1.1 Classification of radioactive waste Radioactive waste shall be classified, collected, and treated separately based on the form of waste generated in medical practice, as well as the types of radionuclides, half-lives, activity levels, and physicochemical properties present.

5.3.1.4.1.2 Waste storage logging and documentation Maintain a waste storage log to record key characteristics and processing procedures, retaining records for archival purposes.

5.3.1.4.1.3 Decay pools and layout Decay pools for radioactive liquid waste should be rationally positioned. Pool bottoms and walls must be robust, resistant to acid/alkali corrosion, impermeable, and equipped with leak-prevention measures.

5.3.1.4.1.4 Dedicated toilets and cleaning Medical institutions conducting radiopharmaceutical therapy shall provide dedicated toilets with protective signage for inpatients or examinees. These toilets shall be designed to ensure all waste is flushed directly into the radioactive liquid decay pool and maintained in a clean state at all times.

5.3.1.4.1.5 Waste containers and markings Waste collection containers shall feature an outer protective layer and ionizing radiation warning labels.

5.3.1.4.2 Sealed waste transport and storage management

5.3.1.4.2.1 Placement of waste containers Place waste containers in locations such as injection rooms, postinjection patient waiting areas, and medication administration rooms.

5.3.1.4.2.2 Sharps containers and handling Radioactive waste containing sharp objects or sharp edges, such as syringes and broken glassware, should first be placed in sharps containers before being placed in dedicated plastic bags.

5.3.1.4.2.3 Waste containers and labeling Waste bags, waste bins, and other storage containers must be secure and reliable. Clearly label each container with waste type, nuclide category, storage date, and other relevant information.

5.3.1.4.2.4 Waste bag handling Place dedicated plastic bags inside waste bins to directly collect waste. Once filled, seal waste bags securely to prevent leakage. Promptly transfer sealed bags to the storage room and place them in designated containers.

5.3.1.4.2.5 Waste dose rate and mass standards The surface dose rate of each waste bag shall not exceed 0.1 mSv/h, and the mass shall not exceed 20 kg.

5.3.1.4.2.6 Waste package surface contamination control standard The contamination control level for the outer surface of waste packages: $\beta < 0.4$ Bq/cm^2.

5.3.1.4.2.7 Storage facility standards Storage areas shall be equipped with ventilation facilities and display ionizing radiation warning signs at entry/exit points.

5.3.2 Professional competency and training for personnel in indirectly nuclear-related industries

5.3.2.1 The value of talent development

5.3.2.1.1 A critical strategic issue for China's nuclear power industry

Talent is a resource more vital than capital or material assets—it is the primary resource [6]. The ability to cultivate and sustain a large-scale, high-caliber talent pool is a critical strategic issue for China's nuclear power industry [19].

5.3.2.1.2 Prerequisite and foundation for nuclear resource development and rational utilization

The continuous advancement of personnel knowledge and skills is essential for fully unlocking the potential of nuclear energy resources. Breakthroughs and developments in science and technology form the prerequisite and foundation for the development and rational utilization of nuclear resources. Therefore nuclear science and technology professionals play a pivotal role in the advancement of nuclear energy.

5.3.2.1.3 Foundational role in advancing nuclear power development

Professionals in nuclear engineering and technology exert direct influence on advancing nuclear power development, serving as the driving force and a key factor in propelling the growth of the nuclear power industry.

5.3.2.2 Positioning and objectives of talent development

5.3.2.2.1 Positioning

The primary distinction between training students in nuclear engineering and nuclear technology and other disciplines lies in this program's emphasis on cultivating students' "research-oriented," "practical," "technical," and "operational" competencies.

5.3.2.2.2 Objectives

Its core mission is to develop applied research professionals for frontline design, research, and management roles within nuclear power enterprises or research institutions.

5.3.2.3 Characteristics and requirements for nuclear faculty

5.3.2.3.1 Characteristics

The demands of this discipline define its faculty distinctiveness from other fields, requiring instructors to integrate theoretical expertise with practical competence. In other words, faculty must not only master specialized theoretical knowledge but also possess extensive practical experience and skills—referred to as "dual-qualified" instructors.

5.3.2.3.2 Requirements

For nuclear engineering and nuclear technology programs, dual-qualified instructors are particularly crucial. Educational authorities and institutional leadership should prioritize building a robust team of such instructors.

5.3.2.4 Establishment of specialized nuclear engineering laboratory system

5.3.2.4.1 Requirements

Fundamental course experiments for nuclear engineering and nuclear technology majors must be conducted in corresponding laboratories. Professional laboratory instruments and equipment should meet the experimental requirements of required specialized courses. Generally the total fixed assets of teaching laboratories should reach 10,000 yuan per student; laboratory space should reach 1 m^2 per student; and laboratories should continuously update experimental equipment to create modern laboratory conditions and install specialized auxiliary experimental equipment.

5.3.2.4.2 Types

To cultivate students for nuclear power research and technical roles, institutions should generally establish laboratories including Nuclear Power Plant Control and Simulation Lab, Nuclear Power Plant Safety Engineering Lab, Nuclear Power Plant Equipment VR Lab, Nuclear Radiation Protection and Detection Lab, Nuclear Power Plant Thermal Engineering Lab, and Nuclear Power Plant Design Engineering Practice Center.

5.3.2.4.3 Examples

Establishing laboratories provides students with means to bridge theory and practice. Renowned universities' nuclear engineering departments all maintain relevant laboratories. Examples include Overseas institutions: MIT, Berkeley, Michigan, Purdue, Georgia Tech, etc.; and Domestic institutions: Tsinghua University, Xi'an Jiaotong University, Southeast University, etc.

5.4 The impact of nuclear technology on residents living near power plants

5.4.1 Physical and mental well-being

5.4.1.1 Mental health

5.4.1.1.1 Public fear of nuclear facilities

5.4.1.1.1.1 Innate fear Historically the devastating consequences of nuclear weapons and accidents have instilled an innate fear of nuclear facilities in people. Although the incidence rate of nuclear power plant accidents is currently kept extremely low, it remains difficult to alleviate the public's inherent apprehension toward such facilities.

5.4.1.1.1.2 Easy to generate "neighbor avoidance effect" Nuclear power plants exhibit significant negative externalities, making them highly susceptible to NIMBY effects [8,9]. This occurs when the economic benefits accrue to the broader public, whereas the negative impacts—such as degraded living environments and economic losses—fall disproportionately on residents surrounding the facility.

5.4.1.1.1.3 Prone to triggering generalized effects This can disrupt social harmony, expose imbalances in spatial benefit distribution, and create psychological barriers among the public. Such issues readily trigger generalized effects, leading to resistance from affected residents and even conflicts. This significantly hinders public facility construction, severely impacting government operations, social stability, and economic efficiency.

5.4.1.1.2 Public psychological barriers

5.4.1.1.2.1 Perception, emotion, or impression psychology As society's demand for clean energy continues to rise, nuclear power has encountered significant development opportunities. However, public fear and resistance toward nuclear energy development remain the primary obstacles to its advancement. Such sentiments are unlikely to change or dissipate over time or through simple policy incentives. From the perspective of risk perception, there exists a marked disparity between the general public and nuclear power experts regarding their perception of nuclear risks. Experts base their risk assessments primarily on professional data metrics, whereas laypeople perceive risks and associated harms mainly through factors such as familiarity, lethality, controllability, and fear—paying little attention to technical risk indicators [8,10]. It is difficult for the public to quickly weigh risks against benefits by estimating probability and severity; instead, they make decisions based on feelings, emotions, or impressions.

5.4.1.1.2.2 Conformity and group-following psychology Compared to governments and corporations the public perceives their individual risk prevention capabilities as weaker and more vulnerable to harm. Consequently, individuals seek out groups sharing common interests, losses, emotional distress, grievances, and rights demands, actively participating in collective activities. Within such groups, individuals exhibit pronounced conformity, leading to collective behavior characterized by herd mentality and groupthink. This can escalate into a collective psychological crisis. Compared to government or expert statements the behavioral choices of peers within a group exert a stronger influence on residents near nuclear projects.

5.4.1.2 Physical health

5.4.1.2.1 Nuclear power plant operation and safety assurance

Compared to mental health, physical health is a factor that directly impacts nuclear power development. The immense hazards of radioactive nuclides can severely affect human health. However, with the continuous advancement of nuclear technology, all types of nuclear power plants now feature extremely high safety standards. The probability of all four conventional safety barriers failing simultaneously is extremely low. As long as they do not fail concurrently, radioactive fission products within the reactor cannot leak into the surrounding environment.

5.4.1.2.2 Nuclear plant emissions and health doses

Regulations governing potential pollution from nuclear plant operations exist in all nations. Examples include the following:

- Annual radiation dose limits for surrounding residents during normal plant operation and
- Dose limits for radionuclides discharged in plant effluent.

Within certain limits the human body possesses natural resistance and recovery capabilities against radiation damage. The human body can tolerate a single exposure of 0.25 Sv without suffering damage. Therefore nuclear power plants operating normally have virtually no impact on the health of surrounding residents.

5.4.1.2.3 Nuclear power plant impacts and health assessment

Nuclear power plant operations involve multifaceted impacts, particularly those influenced by human subjective operational factors. In addition, compared to traditional power generation equipment, nuclear power has a shorter operational history, resulting in fewer accumulated case studies and management experiences. The operation of nuclear power relies on the long-term, proper functioning of monitoring systems. It is essential to conduct preoperational health surveys for residents near nuclear power plants and to investigate and assess surrounding agricultural and livestock products.

5.4.2 Living environment

5.4.2.1 Aquatic radiation environment

The level of radioactivity around nuclear power plants is related to the receiving water body. Generally speaking the stronger the interaction with the receiving water body, the closer the radionuclide concentration levels in the surrounding waters will be to background levels. For coastal nuclear power plants, seawater exchange is strong. When liquid radioactive waste is discharged into seawater, it rapidly disperses and dilutes, with nuclide concentrations fluctuating within the background range. For inland nuclear power plants, lakes and rivers possess relatively weaker dilution capacity. When discharge volumes are substantial, radionuclides struggle to disperse rapidly. However, both the average dose rate and maximum dose rate remain well below the 10 mGy/d threshold, indicating that radionuclides in nuclear discharge water do not cause radiation damage to aquatic biological communities.

5.4.2.2 *Aquatic thermal environment*

Thermal effects cause an increase in warm-water algae and fish populations around nuclear power plants, leading them to congregate in surface waters and near discharge outlets. These populations are also influenced by water interactions. As the dilution capacity of the water body diminishes, the effects of thermal impact gradually become more pronounced. The numbers of poikilothermic fish and algae gradually decrease, eventually disappearing, whereas warm-water species gain dominance and gradually become the main component of the community.

5.4.2.3 *Atmospheric environment*

In terms of carbon emissions per unit of electricity generated, nuclear power's emissions primarily stem from fuel extraction and waste disposal. This implies that nuclear power plants' contribution to atmospheric pollution is virtually negligible [20].

5.4.2.4 *Soil environment*

Nuclear technology has dual impacts on soil environments. Positively, it enables soil composition analysis, pollution detection, and pest control—such as precise measurement of soil elements, tracking pollutants, and reducing pesticide contamination. Negatively, radioactive materials from nuclear facilities, if improperly managed, can contaminate soil, disrupt soil ecosystems, and potentially harm human health through the food chain. Research by Weichao et al. analyzed surface soils from China's coastal regions for plutonium isotopes (^{239}Pu and ^{240}Pu). Findings indicate that plutonium isotopes in these soils primarily originate from global atmospheric fallout of nuclear weapons tests conducted between 1945 and 1980. Contributions from other human nuclear activities to coastal regions are negligible [16].

5.4.2.5 *Waste treatment*

Nuclear power plant effluents (waste liquids and gases) undergo stringent treatment before discharge. Radioactive waste liquids are processed through evaporation, ion exchange, coagulation-sedimentation, and filtration to meet regulatory standards before being discharged via tanks. Concentrated liquids and high-level radioactive waste liquids are solidified and stored after concentration. Radioactive waste gases undergo filtration, storage, and decay processes; once their radioactivity levels reach permissible values, they are released into the atmosphere via chimneys.

5.4.2.6 Environmental effects

The environmental impact of nuclear power plants is dual-natured, characterized overall by controllable low-level effects while harboring potential risks. During normal operation, stringent purification processes ensure only trace amounts of radioactive substances are released, with doses far below international safety standards, posing minimal impact on the environment and human health. Warm discharge from cooling systems may cause slight local water temperature increases, potentially affecting short-term distribution of aquatic life in the area. This can be mitigated through optimized discharge design and cooling device installation. Furthermore, as a clean energy source, nuclear power plants significantly reduce emissions of greenhouse gases such as carbon dioxide and sulfur dioxide, as well as other pollutants, contributing to climate change mitigation.

5.4.3 Socioeconomic benefits

5.4.3.1 Economic benefits

Nuclear power plants are typically located far from residential areas, often limiting regional development. The presence of a nuclear power plant can significantly boost the local economy [11,12]. During construction a large workforce will be stationed on-site, with the plant's permanent staff reaching tens of thousands. The extended construction period creates a sustained consumer base, substantially driving local spending on dining and daily living expenses. During operation, beyond plant owners, maintenance and equipment operation personnel also reside nearby. This fosters residential clusters around the plant, where diverse living needs further stimulate local industrial growth.

5.4.3.2 Energy supply benefits

Nuclear power plants also assist in adjusting local energy structures. For regions requiring heating the heat supply from conventional power plants' combined heat and power (CHP) systems is extremely limited in coverage, confining most heating to local areas. However, a single nuclear power unit can provide heating for a vast residential area, significantly expanding the heating coverage zone. Economically, nuclear heating (including district heating and steam supply) costs are comparable to coal-fired heating and offer advantages over gas heating.

5.5 The impact of nuclear technology on the public

5.5.1 Safety and radiation

5.5.1.1 Impact of nuclear safety and reliability

In today's world centered on peace and development, nuclear safety and radiation safety hold paramount importance. They are crucial to the healthy and sustainable development of human society and civilization, as well as to the prosperity and advancement of nations and peoples. Nuclear radiation can cause immense damage and impact on both human health and the environment. Therefore, from the very outset of nuclear energy development, nuclear safety has been accorded the highest priority. Fundamental principles of nuclear safety were established early on, with particular emphasis placed on ensuring safety through reliability in design and equipment. The level of nuclear safety is the most critical factor determining public acceptance of comprehensive nuclear energy utilization.

5.5.1.2 *Nuclear safety and radiation protection objectives*

According to the 2006 regulations of the IAEA the primary objective of nuclear safety has been, since the early stages of civil nuclear fission technology, to protect humans and the environment from the hazards of ionizing radiation. Achieving this objective is essential to ensure the normal operation of nuclear facilities. Therefore controlling personnel radiation exposure and the release of radioactive materials into the environment is critical under both normal and abnormal operating conditions. Mitigating radiation consequences during unexpected events is also essential. To this end, governments, facility owners, and other relevant organizations must collaborate and implement necessary actions to ensure nuclear safety and radiation protection.

5.5.1.3 *Impact of nuclear power plants on the public*

Nuclear power plants are renowned for their long-term reliable operation, low greenhouse gas emissions, and high power density. Nevertheless, concerns persist regarding the potential health and environmental impacts of their radioactive emissions. Kumar et al. conducted a comprehensive analysis of radiation data collected over 20 years (2000–20) from seven Indian nuclear power plants to assess their radioactive impact on the environment and public health [14]. The analysis yielded several key conclusions: radioactive concentrations around the plants were low, comparable to measurements at other nuclear power plants worldwide; tritium concentrations in receiving water bodies were less than 1% of the internationally recommended limit of 10,000 Bq/L; estimated total radiation doses to the public were at most 10% of the prescribed 1000 μSv regulatory dose limit and showed a consistent decline over the study period. The findings indicate that the radioactive impact of nuclear power plants on the environment and public is negligible.

5.5.1.4 *Impact of nuclear technology on the public*

Nuclear technology plays a vital role in medical diagnosis and treatment. Techniques such as nuclear imaging [including positron emission tomography (PET) and single-photon emission computed tomography (SPECT)] enable physicians to precisely detect and diagnose diseases like cancer. In addition, nuclear medicine therapies, such as radiation therapy, provide targeted treatment options for various conditions. Nuclear technology finds applications in diverse industrial processes, including sterilization, food preservation, and material testing. Radiation techniques enhance product safety, quality, and efficiency across industries. In agriculture, nuclear technology supports crop improvement, pest and disease control, and soil management. Radiation-based techniques aid crop breeding, yielding disease-resistant and high-yielding varieties. Furthermore, nuclear applications contribute to pest eradication and food irradiation, extending shelf life and ensuring food safety.

5.5.1.5 Public impact of nuclear waste and nuclear security

One of the primary challenges associated with nuclear technology is the management of radioactive waste. The disposal of nuclear waste poses risks to the environment and public health, necessitating stringent safety measures and long-term storage solutions. Ensuring the safety and security of nuclear facilities is critical to protecting public health and the environment. Rigorous regulations, robust safety protocols, and international cooperation are essential to prevent accidents involving radioactive materials, nuclear proliferation, and malicious acts.

5.5.2 Economy and employment

5.5.2.1 The role of nuclear energy in energy structure adjustment

5.5.2.1.1 Reducing carbon emissions

Nuclear energy plays a vital role in energy structure adjustment by providing a reliable, low-carbon source of electricity. It helps reduce carbon emissions, thereby facilitating the transition to cleaner energy systems.

5.5.2.1.2 Reducing reliance on fossil fuels

Nuclear energy diversifies the energy mix, reducing dependence on fossil fuels and enhancing energy security. Fossil fuels may be subject to geopolitical uncertainties.

5.5.2.1.3 Providing reliable baseload power

The stability and reliability of nuclear power within the electricity grid are particularly crucial. Nuclear power plants deliver steady baseload electricity, complementing intermittent renewable sources such as wind and solar.

5.5.2.1.4 Sustainability in energy transition

Investing in nuclear infrastructure expands energy capacity to meet growing electricity demand in a sustainable manner. The role of nuclear energy in energy mix adjustments aligns with the sustainability, security, and reliability objectives of the global energy transition.

5.5.2.2 Impact of nuclear technology on human socioeconomic and environmental security

5.5.2.2.1 Diversifying energy sources to enhance energy security

Beyond energy production, nuclear technology has applications in medicine, agriculture, and manufacturing, further driving economic development. Nuclear energy diversifies energy sources, reduces reliance on fossil fuels, and thereby strengthens energy security [11,12].

5.5.2.2.2 Stable and sustained energy supply promotes industrial growth and social stability

Increasing the number of countries with nuclear capabilities can mitigate risks stemming from energy market volatility and geopolitical tensions. Furthermore, nuclear energy provides a stable and continuous energy supply, which is crucial for industrial growth and social stability.

5.5.2.2.3 Reducing air pollution and advancing environmental protection

Nuclear energy serves as a low-carbon alternative to fossil fuels, aiding in climate change mitigation. Unlike coal or natural gas power plants, nuclear facilities emit minimal greenhouse gases during operation, helping to reduce air pollution and lessen the ecological footprint of electricity generation [22].

5.5.2.2.4 Minimizing radiation and waste to enhance environmental credibility

Advancements in nuclear technology aim to address challenges in radioactive waste management and reactor safety, thereby further improving its environmental credibility.

5.5.2.3 Nuclear science and technology innovation and development

5.5.2.3.1 Advancing healthcare development

Nuclear science and technology play a vital role in healthcare, advancing medical diagnosis, treatment, and research. Technologies, such as nuclear imaging, radiation therapy, and nuclear medicine, have revolutionized medical practice, enabling early disease detection, precision treatment, and innovative drug development. In addition, nuclear technology supports advancements in biotechnology, genetics, and pharmaceuticals, contributing to improved medical outcomes and disease management.

5.5.2.3.2 Advancing agricultural development

Technology enhances agricultural productivity, ensures food security, and supports rural livelihoods. Techniques such as radiation breeding and isotope tracing increase crop yields, improve soil fertility, and mitigate pest damage, thereby addressing global food supply challenges [15].

5.5.2.3.3 Improving food quality

Applications of nuclear technology in food preservation, irradiation, and quality control extend shelf life, reduce food waste, and maintain food safety standards.

5.5.2.3.4 Stimulating economic growth and industrial advancement

Nuclear technology drives innovation across industries, fostering technological progress and economic competitiveness. Derivative technologies from nuclear physics, materials science, and engineering research find applications in telecommunications, computing, and materials manufacturing, thereby stimulating economic growth and industrial advancement. Nuclear technology also supports the development of advanced materials, nanotechnology, quantum computing, and AI, propelling innovation and shaping the future of industry.

5.5.2.4 Nuclear technology drives diverse employment needs

5.5.2.4.1 Demand for skilled labor

Nuclear technology has profoundly impacted humanity's socioeconomic landscape, presenting both opportunities and challenges. Nuclear energy promotes economic growth by providing reliable and affordable electricity, supporting diverse industries, and creating jobs. The construction, operation, and maintenance of nuclear power plants require a technically skilled workforce [5,6], thereby generating high-quality, high-income employment opportunities. From energy production to healthcare and engineering design, nuclear technology significantly impacts job creation across industries, offering numerous employment opportunities and positions. Operating and maintaining nuclear power plants requires a skilled workforce, including nuclear engineers, reactor operators, technicians, and maintenance personnel. These professionals ensure safe and efficient power generation, contributing to grid stability and meeting community energy needs.

5.5.2.4.2 Specific job types

5.5.2.4.2.1 Radiation protection specialist In nuclear facilities and medical settings, radiation protection specialists play a critical role in ensuring worker and public safety. They monitor radiation levels, enforce safety protocols, and provide training in radiation safety practices.

5.5.2.4.2.2 Nuclear engineers Nuclear engineers design, develop, and operate nuclear systems and equipment. They engage in reactor design, nuclear fuel cycle management, waste treatment, and radiation shielding projects.

5.5.2.4.2.3 Health physicists Health physicists specialize in radiation protection for medical, industrial, and environmental applications. They assess radiation risks, establish safety procedures, and ensure compliance with regulatory standards.

5.5.2.5 Talent demand in nuclear technology

5.5.2.5.1 Nuclear medicine technicians

In healthcare, nuclear medicine technicians operate imaging equipment to create diagnostic images and administer radiopharmaceuticals for medical procedures such as PET scans and nuclear stress tests.

5.5.2.5.2 Research scientists

Research scientists conduct experiments, analyze data, and develop new technologies for energy, medicine, and industry. With increasing emphasis on nuclear waste disposal and remediation, waste management specialists are needed to devise safe storage solutions, implement decontamination procedures, and monitor environmental impacts. Nuclear technology drives research and innovation in fields such as nuclear physics, materials science, and nuclear medicine.

5.5.2.5.3 Instrumentation and control technicians

Instrumentation and control technicians install, calibrate, and maintain instruments and control systems within nuclear facilities. They ensure equipment operates properly and monitor critical parameters to guarantee safety and efficiency.

5.5.2.5.4 Regulatory affairs specialists

Regulatory affairs specialists ensure compliance with nuclear regulations and licensing requirements. They liaise with regulatory bodies, prepare documentation, and assist with inspections to uphold safety standards.

5.5.2.5.5 Educators and trainers

As the nuclear industry continues to evolve, there is a growing need for educators and trainers to cultivate the next generation of nuclear professionals. These individuals develop curricula, implement training programs, and mentor students and new employees.

5.5.2.6 Evaluation

Nuclear technology has profound socioeconomic impacts, ranging from energy security and environmental sustainability to healthcare and industrial innovation. From power generation to healthcare and research, nuclear-related industries offer diverse employment opportunities for individuals with varied skills and backgrounds. Although nuclear energy delivers substantial benefits for economic growth, job creation, and environmental stewardship, it also presents challenges such as proliferation risks, safety concerns, and radioactive waste management. Therefore harnessing the potential of nuclear technology requires careful consideration of its socioeconomic impacts alongside robust regulatory frameworks, international cooperation, and continuous technological innovation.

5.5.3 Anxiety and pride

5.5.3.1 Negative psychological impacts of nuclear technology

5.5.3.1.1 Psychological impact of nuclear war threats and radiation

5.5.3.1.1.1 Inducing feelings of powerlessness and despair The threat of nuclear war and the risks of radiation pose immense psychological challenges to humanity and society worldwide. Merely the threat of nuclear war or the possibility of radiation exposure can trigger profound fear and anxiety among the public. The existential threat posed by nuclear weapons and warfare triggers primal survival instincts, intensifying stress responses and psychological distress. The unpredictability of nuclear incidents and their catastrophic consequences foster widespread uncertainty and helplessness. Individuals' inability to control or mitigate the impact of nuclear threats generates feelings of powerlessness and despair [6,9].

5.5.3.1.1.2 Causing acute trauma and posttraumatic stress disorder (PTSD) The atomic bombings of Hiroshima and Nagasaki in August 1945 inflicted unprecedented destruction and psychological trauma on local residents [18]. They experienced intrusive memories, flashbacks, and nightmares of the traumatic event, repeatedly reliving the horror of the atomic explosions. Persistent excessive anxiety and hypervigilance further deteriorated their mental health and quality of life. Direct exposure to nuclear warfare or radiation incidents may result in acute trauma and posttraumatic stress disorder (PTSD).

5.5.3.1.2 Psychological anxiety from nuclear accidents and leaks

5.5.3.1.2.1 Exacerbating psychological impacts of trauma and loss Nuclear threats and accidents often trigger widespread panic and social chaos, manifesting as mass evacuations, stockpiling of supplies, and breakdowns in social order. Collective fear amplifies individual anxiety, leading to social unrest and instability. Communities affected by nuclear incidents may face stigmatization and discrimination, intensifying the psychological impacts of trauma and loss.

5.5.3.1.2.2 Inducing guilt, grief, and despair Fear-based prejudice and stigma further isolate individuals, hinder recovery efforts, and exacerbate mental health challenges. The environmental consequences of nuclear accidents intensify ecological anxiety and environmental distress. Concerns over radioactive contamination, ecosystem destruction, and long-term health impacts trigger feelings of environmental guilt, grief, and despair, particularly among younger generations.

5.5.3.1.3 Evaluation

The threat of nuclear war and the risk of nuclear accident leaks inflict profound negative psychological impacts on individuals and society. Fear, anxiety, uncertainty, and trauma permeate entire communities, undermining people's mental health and well-being.

5.5.3.2 *The positive psychological impact of nuclear technology*

5.5.3.2.1 The influence of nuclear technology advancements

5.5.3.2.1.1 Stimulating curiosity, excitement, and hope Nuclear technology has made profound contributions to society, evident not only in technological progress but also in shaping people's psychological outlook. Nuclear technology has consistently served as a catalyst for innovation, driving progress across diverse fields. The pursuit of harnessing nuclear energy for power generation has propelled advancements in physics, engineering, and materials science. The quest for scientific discoveries and technological breakthroughs ignites curiosity, excitement, and hope for the future, thereby promoting psychological well-being [17].

5.5.3.2.1.2 Instilling a sense of responsibility and pride The development of nuclear energy offers sustainable solutions to global energy crises while mitigating the environmental impact of fossil fuels. The shift toward clean energy reduces air pollution, greenhouse gas emissions, and dependence on finite resources, fostering environmental responsibility and pride in contributing to a healthier planet.

5.5.3.2.1.3 Providing security and empowerment Nuclear technology has revolutionized medical diagnostics and treatments, enhancing healthcare outcomes and improving quality of life. Diagnostic imaging techniques such as PET and magnetic resonance imaging (MRI) enable early detection and precise diagnosis of diseases, alleviating patient anxiety and improving treatment effectiveness. Access to advanced healthcare delivers a sense of security and empowerment, fostering a positive and optimistic mindset.

5.5.3.2.1.4 Fostering solidarity and trust The collaborative nature of nuclear research promotes international partnerships and scientific exchange, deepening cross-cultural understanding and friendly relations among nations. The shared goals of scientific exploration and technological advancement transcend geopolitical boundaries, cultivating a sense of solidarity and trust in humanity's collective capacity to address global challenges.

5.5.3.2.2 Positive impacts of nuclear technology

5.5.3.2.2.1 Restoring trust in nuclear technology Despite devastating nuclear accidents at Chernobyl and Fukushima, affected regions demonstrated remarkable resilience and strength in adversity. Psychological support programs, regional recovery initiatives, and infrastructure reconstruction efforts played a crucial role in restoring trust in nuclear technology.

5.5.3.2.2.2 Fostering optimism and gratitude Nuclear medicine has revolutionized the diagnosis and treatment of various diseases, bringing hope and healing to patients worldwide. From cancer therapy to cardiac imaging, nuclear technology has saved countless lives, alleviated suffering, and instilled optimism and gratitude among patients, caregivers, and healthcare professionals.

5.5.3.2.2.3 Fostering pride and confidence The integration of nuclear power with renewable sources such as wind and solar makes energy infrastructure more resilient and sustainable. Regions with clean, reliable energy solutions experience enhanced economic stability, social cohesion, and environmental stewardship, cultivating a sense of pride and confidence in their ability to shape a brighter future.

5.5.3.2.3 Evaluation

The advancement of nuclear technology has profoundly and multifacetedly impacted societal progress through positive psychological effects. From fostering innovation and progress to driving environmental stewardship and improving healthcare outcomes, nuclear technology's evolution has ignited hope, resilience, and collaboration worldwide. By harnessing the transformative power of nuclear technology, humanity can continue to embrace a brighter, more sustainable future for generations to come.

5.5.3.3 Strategies for addressing nuclear risks

5.5.3.3.1 Promoting education and awareness of nuclear threats and risks

Enhancing public education and awareness of nuclear threats and risks empowers individuals to make informed decisions and take proactive measures to protect themselves [21]. Knowledge dissemination strengthens resilience and adaptive coping strategies in the face of uncertainty.

5.5.3.3.2 Support collective action and build resilient nations and regions

Government support and collective action to build resilient nations and regions can improve mental health and foster a sense of solidarity. Strengthening mutual aid, emotional support, and cohesion enhances public resilience and facilitates recovery from nuclear-related trauma. Advocating evidence-based policies and risk mitigation strategies is crucial for addressing nuclear threats and minimizing their psychological impact.

5.5.3.3.3 Investing in preparedness, crisis management, and nuclear nonproliferation

Investing in preparedness, crisis management, and nuclear nonproliferation efforts helps restore public confidence and reduce anxiety.

5.5.3.3.4 Strengthening education, government support, and policy advocacy

Through strengthened education, government support, and policy advocacy, individuals and societies can confront nuclear challenges with resilience and hope, creating a safer, more secure world for future generations.

References

[1] A. Mardha, A. Mardhi, T. Asmoro, et al., Nuclear security policy challenges and strategies in the Serpong nuclear area: a regulatory review, AIP Conf. Proc. 2967 (1) (2024) 0194397.

[2] B. Jing, A study of China-Latin America cooperation in green energy industry: status, obstacles, and the policy suggestions, Int. J. Sustain. Dev. Plan. 19 (6) (2024), doi:10.18280/ijsdp.190636.

[3] F.S.M. Chachuli, A. Aziz, F.M. Idris, Technology foresight: capturing the benefits from nuclear science technology and innovation in Malaysian Nuclear Agency, IOP Conf. Ser.: Mater. Sci. Eng. 1308 (2024) 12008, doi:10.1088/1757-899X/1308/1/012008.

[4] M. Amiri, S.N. Mousavi, R. Moghaddasi, The role of development and financial risk in agricultural energy consumption: an empirical analysis based on country panel data and smooth transition regression, Braz. J. Biol. (2024) 84, doi:10.1590/1519-6984.288693.

[5] W. Edward, A. Kirk, W. Anthony, A perspective from Ontario Tech University industrial research chairs on 20 years of capacity building in health physics and radiation science, Health Phys. 126 (5) (2024) 280–291.

[6] H.M. Jayathilake, L.E. Botsyoe, H. Said, The impact of it evolution on industries and workforce skills: a systematic literature review, Issues Inf. Sci. Inf. Technol. 21 (2024) 5327.

[7] Q. Ma, Z. Chen, J. Chen, et al., A study on the evolution of flame height and air entrainment volume rate for typical electrical cabinet fires in nuclear power plants, Appl. Energy 382 (2025) 125338.

[8] R. Mills, E.M. Merz, M. Croucher, et al., The infected blood inquiry: impact on public perceptions of blood supply risk, safety, and donation attitudes, Transfus. Med. 34 (6) (2024) 13108.

[9] Y. Wang, J. Gu, J. Wu, Explaining local residents' acceptance of rebuilding nuclear power plants: the roles of perceived general benefit and perceived local benefit, Energy Policy (2020) 140, doi:10.1016/j.enpol.2020.111410.

[10] A. Creulescu, Small modular reactors in Romania's energy future. Two essential perspectives: capital costs and public perception, Proc. Int. Conf. Bus. Excell. 18 (1) (2024) 1765–1775, doi:10.2478/picbe-2024-0148.

[11] I. Vakulenko, O. Lyulyov, T. Pimonenko, et al., The impact of energy on a country's social and economic development, AIP Conf. Proc. 3033 (1) (2024) 14, doi:10.1063/5.0189046.

[12] O.F. Shaqqour, The Impact of Integrating Strategic Management Accounting and Decision-Making on Financial Performance in Industrial Companies, Springer, Cham, 2025, doi:10.1007/978-3-031-74220-0_22.

[13] M. Sáez-Muñoz, A. Cerezo, E. Prieto, M. Salvadó, I. Vildosola Hernandez, M.A. Duch, A. Camp, E. Gallego, J. Gonzalez-Cadelo, G. Verdú, Recent radiation protection activities related to nuclear facilities on the Iberian Peninsula, Nucl. Eng. Des. 417 (2024) 112826.

[14] A.V. Kumar, A.K. Patra, S.N. Tiwari, A. Baburajan, Y.P. Gautam, B. Vijayakumar, T. Jesan, M.S. Vishnu, I.V. Saradhi, A. Chandra, D.K. Aswal, Negligible radiological impact of Indian nuclear power plants on the environment and the public: findings from a 20-year study, Sci. Total Environ. 914 (2024) 169936.

[15] E. Ebrahimi Sarcheshmeh, M. Bijani, H. Sadighi, Adoption behavior towards the use of nuclear technology in agriculture: a causal analysis, Technol. Soc. 55 (2018) 175–182.

[16] IEANuclear Power and Secure Energy Transitions, IEA, Paris, 2022.

[17] D.B. Yaden, J.C. Eichstaedt, J.D. Medaglia, The future of technology in positive psychology: methodological advances in the science of well-being, Front. Psychol. 9 (2018) 962.

[18] R.J. Lifton, Psychological effects of the atomic bomb in Hiroshima: the theme of death, Daedalus 92 (3) (1963) 462–497.

[19] K. Al-Tarawneh, The impact of human resource management strategies on the strategic success of cellular communications companies in Jordan, Int. J. Serv. Sci. Manag. Eng. Technol. 14 (1) (2023) 1–23.

[20] T. Koljonen, I. Savolainen, Impact of the eu emissions trading directive on the energy and steel industries in Finland, in: Greenhouse Gas Control Technologies 7: Proceedings of the 7th International Conference on Greenhouse Gas: Control Technologies 5 September 2004, Vancouver, Canada, 1, 2005, pp. 833–840.

[21] J.A. Domenech, R.M. Clary, Student and public perceptions of stormwater runoff and its impact on public health in a Southern California coastal community, J. Geosci. Educ. 73 (1) (2025) 2366695.

[22] S. Rauner, F. Piontek, B. Soergel, et al., The impact of energy sector pollution on human development and inequality amidst climate change, Environ. Res. Lett. 19 (2024) 094042, doi:10.1088/1748-9326/ad6b39.

Chapter 6

The environment and human demand for nuclear technology

Tao Zhou[a], Shang Mao[b], Dong Wei[a], Huaichang Lu[a] and Tianyu Gao[a]
[a] *Department of Nuclear Science and Technology, Institute of Nuclear Thermal-hydraulic Safety and Standardization, Southeast University, Nanjing, Jiangsu, China,* [b] *School of Energy and Power Engineering, Zhengzhou University of Light Industry, Zhengzhou, Henan, China*

6.1 Humanity's demand for nuclear technology

6.1.1 Human understanding of nuclear technology

6.1.1.1 Humanity's journey into nuclear science

6.1.1.1.1 Foundational understanding of atomic structure

Humanity's exploration of nuclear science began in the early 20th century. With advances in atomic structure theory and experimental techniques, scientists started uncovering the mysteries within atoms. In 1911 Ernest Rutherford proposed the planetary model of the atom, revealing for the first time the existence of a unified atomic nucleus at the core—the seed of nuclear science. Subsequently, James Chadwick discovered the neutron in 1932, laying the foundation for later research into nuclear fission and fusion.

6.1.1.1.2 Foundational nuclear experiments

In 1938 Otto Hahn and Fritz Strassmann conducted uranium fission experiments in Berlin, marking the dawn of the nuclear age. The following year, Leo Szilard recognized the potential for chain reactions in nuclear fission, which spurred the launch of the Manhattan Project.

6.1.1.1.3 Understanding atomic weapons

Deepening theoretical understanding ultimately led to the successful development of the atomic bomb. In 1945 the United States conducted the world's first nuclear explosion test at Trinity Site in New Mexico. That same year, atomic bombs were dropped on Hiroshima and Nagasaki, Japan, shocking the world with the destructive power of nuclear weapons.

Nuclear Science, Technology, and Civilization. DOI: https://doi.org/10.1016/B978-0-443-13941-3.00006-9
Copyright © 2026 Elsevier Inc. All rights are reserved, including those for text and data mining, AI training, and similar technologies.

6.1.1.1.4 Peaceful applications of nuclear energy

After World War II, humanity began recognizing nuclear energy's peaceful uses. In 1954 the Soviet Union constructed the world's first commercial nuclear power plant—the Obninsk reactor—ushering in a new era of nuclear power generation. The United States utilized nuclear plants to provide stable electricity, apply radiation therapy and medical imaging in cancer treatment, process industrial materials, and produce isotopes for scientific research.

6.1.1.1.5 Reflection and reassessment following severe nuclear incidents

Since then, hundreds of nuclear power plants have been constructed globally, establishing nuclear energy as a vital source of clean power. However, major nuclear accidents such as Chernobyl, Three Mile Island, and Fukushima have prompted profound reflection on nuclear safety.

6.1.1.2 The development, application, and ethical challenges of nuclear technology

6.1.1.2.1 The potential for nuclear technology development and application

As understanding of nuclear technology deepens, humanity has progressively uncovered its immense application potential. In medicine, radioactive isotopes are used to diagnose and treat various diseases, such as radioactive iodine therapy for thyroid cancer and positron emission tomography (PET) scans for tumor diagnosis. Furthermore, nuclear medicine has advanced the development of biomarkers and molecular imaging (MI) techniques, providing powerful tools for precision medicine. Nuclear technology also plays a vital role in materials science, environmental monitoring, archaeology, and geology. Through radiocarbon dating, scientists can precisely determine the age of ancient artifacts and fossils.

6.1.1.2.2 Risks and challenges coexist in nuclear technology development

The utilization of nuclear energy brings challenges, such as the safe disposal of nuclear waste, the risk of nuclear proliferation, and issues of nuclear energy sustainability. The international community strives to control nuclear proliferation through agreements such as the Nuclear Non-Proliferation Treaty while simultaneously developing safer, cleaner nuclear energy technologies. Examples include advanced nuclear technologies and fusion reactor technology—the latter is particularly viewed as a future energy hope because of its theoretical near-zero greenhouse gas emissions and abundant fuel resources [17].

6.1.1.2.3 Balancing technological advancement and ethical responsibility

Human understanding of nuclear technology has evolved through a process of continuous deepening. From the initial exploration of atomic structure to today's peaceful utilization of nuclear energy and stringent nuclear safety management, nuclear technology has not only propelled scientific progress but also compelled society to continually reflect on how to balance technological development with ethical responsibility [1].

6.1.2 Human demand for nuclear technology in industry

6.1.2.1 Nuclear reactor operation and monitoring

People demand safe nuclear energy. Technicians can leverage artificial intelligence (AI) to analyze reactor operational data—such as temperature, pressure, and power parameters—to build models predicting performance changes [18]. Based on these forecasts, they optimize reactor operating conditions to enhance power generation efficiency and reduce energy consumption. For instance, machine learning algorithms optimize core power distribution, maximizing fuel utilization and extending refueling cycles. Leveraging deep learning, AI can monitor reactor status in real time, promptly detecting potential failure indicators. By learning from vast historical fault and normal operation data, AI systems identify anomalous patterns and issue early failure warnings, enabling staff to take corrective actions and prevent accidents. For instance, it can detect abnormal flow changes in cooling systems, anticipating potential coolant leaks [2].

6.1.2.2 Nondestructive testing

Nuclear technology plays a pivotal role in industrial nondestructive testing. Radiography using gamma rays and X-rays inspects internal defects in metal components, welds, and composite materials without damaging the tested objects. This technology is widely applied in aerospace, oil and gas, construction, and other industries to ensure structural integrity and safety.

6.1.2.3 Radiation technology

Radiation technology alters material properties. For instance, radiation crosslinking enhances plastics' heat resistance, abrasion resistance, and chemical stability—commonly used in cable insulation and pipeline linings. In addition, radiation curing instantly solidifies coatings, adhesives, and printing inks, boosting production efficiency while reducing solvent use. Radiation sterilization is a nonthermal method suitable for medical devices, food packaging, and pharmaceuticals that cannot withstand high temperatures. By employing electron beams, X-rays, or gamma rays from sources like cobalt-60, microorganisms can be effectively eliminated, extending product shelf life while preserving physical and chemical properties.

6.1.2.4 Radioisotopes

Radioisotope sources facilitate online thickness measurement, density measurement, and flow monitoring, enabling real-time production adjustments to ensure product compliance with specifications. For example, on paper and film production lines, beta and gamma ray sensors continuously monitor thickness variations, automatically adjusting machine settings to maintain consistent quality standards.

6.1.2.5 Nuclear detection technology

Nuclear logging technology is critical in oil and gas exploration. By measuring natural radioactive element concentrations in rock formations, it assesses underground rock types and properties, guiding drilling operations [19].

6.1.2.6 Neutron activation analysis

Nuclear techniques such as neutron activation analysis and X-ray fluorescence spectroscopy enable precise ore composition analysis, optimizing mining and mineral processing workflows [20].

6.1.2.7 Radioisotope thermoelectric generators (RTGs)

For industrial equipment requiring long-term stable power, such as deep-sea detectors and remote sensor networks, RTGs (radioisotope thermoelectric generators) offer a reliable solution. They harness heat generated by radioactive isotope decay, converting it into electricity, enabling decades of continuous operation in extreme environments.

6.1.2.8 Environmental monitoring

Nuclear technology also plays a vital role in environmental monitoring. Radioactive tracers, for instance, track contaminant migration paths in water bodies and soil, aiding in developing effective remediation strategies. In addition, radioactive isotopes serve as markers to study the sources and distribution of atmospheric particulate matter, supporting air quality improvement measures.

6.1.2.9 Robotics and automated systems

In high-radiation environments, such as nuclear facility maintenance and decommissioning projects, specialized robots and automated systems are essential. These devices are typically equipped with radiation-resistant sensors and actuators, enabling operations in areas inaccessible to humans while enhancing efficiency and safety [3].

6.1.2.10 Evaluation

The application of nuclear technology in industry demonstrates its powerful functionality and versatility, not only enhancing production efficiency and product quality but also solving many challenges that traditional methods struggle to overcome. However, strict radiation safety standards must be followed when using nuclear technology to ensure personnel health and environmental protection. With technological progress and innovation the application prospects of nuclear technology in industrial fields will continue to expand, promising to provide stronger technical support and innovative momentum for the transformation and upgrading of modern industrial systems and high-quality socioeconomic development.

6.1.3 Humanity's demand for nuclear technology in agriculture

6.1.3.1 Radiation mutagenesis

Nuclear technology, especially radiation mutagenesis, has opened up new avenues for crop breeding and improvement. By using gamma rays, X-rays, or fast neutrons to irradiate seeds or plants, genetic mutations can be induced, thereby creating new varieties with better stress resistance, higher yield, or better quality. This method has successfully cultivated excellent varieties of various crops, such as rice, wheat, and soybeans.

6.1.3.2 Pest control

The main application of nuclear technology in pest control is through the release of sterilized insect technology (SIT). This method uses the radiation produced by radioactive isotopes to sterilize male pest individuals and then releases them into the wild to mate with wild females but be unable to reproduce offspring, gradually reducing the pest population. This method is environmentally friendly and efficient and has achieved significant results in controlling pests such as Mediterranean fruit flies and mosquitoes.

6.1.3.3 Pesticide residues

Nuclear technology can help detect pesticide residues and other pollutants, ensuring food safety. For example, using accelerator mass spectrometry (AMS) technology, trace amounts of radiocarbon (C-14) in food can be accurately determined to determine the degradation and residue levels of pesticides.

6.1.3.4 Radiation sterilization

Nuclear technology can also be used for radiation sterilization of food to kill bacteria and viruses, extend the shelf life of food, and preserve its nutritional value.

6.1.3.5 Isotope tracking

By using radioactive isotope tracing technology the dynamics of water and nutrients in the soil–plant system can be tracked. For example, tritium (H-3) and deuterium (D) can be used as markers for water, which can help understand the distribution and utilization efficiency of irrigation water and guide water-saving agricultural practices. In addition, by labeling elements such as nitrogen, phosphorus, and potassium in fertilizers, the ability of plants to absorb nutrients can be evaluated and fertilization strategies can be optimized.

6.1.3.6 Environmental monitoring

Nuclear technology also plays a role in agricultural environmental monitoring. Radioisotopes such as cesium-137 (Cs-137) and lead-210 (Pb-210) serve as indicators in sediment chronology, reconstructing soil erosion histories and assessing the impacts of land management practices. Furthermore, nuclear techniques can monitor the accumulation and transfer of radioactive contaminants within agricultural ecosystems.

6.1.3.7 Hybridization experiments

In genetics and biotechnology, nuclear techniques are employed to construct genetically modified crops and perform gene editing. Radioisotope-labeled probes facilitate DNA hybridization experiments, aiding in the identification of specific gene sequences and advancing marker-assisted breeding. Radioactively labeled proteins and antibodies also find applications in proteomics research, contributing to the understanding of molecular mechanisms underlying crop growth, development, and responses to environmental stress.

6.1.3.8 Evaluation

The application of nuclear technology in agriculture is both a specialized and extensive field, utilizing nuclear techniques to enhance crop yields, improve food quality, control pests and diseases, and monitor environmental pollution. This application demonstrates the interdisciplinary nature of nuclear technology, integrating knowledge from physics, chemistry, biology, and environmental science. These technologies not only boost agricultural productivity and food quality but also contribute to environmental protection and human health. However, the use of radioactive materials necessitates strict adherence to radiation safety regulations, ensuring all operations occur under secure and controlled conditions. As agricultural technology continues to advance, the application of nuclear technology in agriculture is poised for further expansion, poised to make greater contributions toward achieving sustainable agriculture.

6.1.4 Humanity's demand for nuclear technology in military applications

6.1.4.1 *Nuclear weapons*

The advent of nuclear weapons represents the most direct manifestation of nuclear technology's military application. In 1945 the United States successfully developed and tested the world's first atomic bomb under the Manhattan Project. Its subsequent deployment over Hiroshima and Nagasaki shocked the world, heralding the dawn of the nuclear age. Nuclear weapons fall into two main categories: fission and fusion weapons (hydrogen bombs). The former harnesses the energy released by a chain reaction of fission in heavy nuclei (such as ^{235}U or ^{239}Pu), whereas the latter utilizes the fusion reaction of lighter nuclei (such as deuterium and tritium) to produce a greater energy output. Although possessing immense destructive power, nuclear weapons also bring catastrophic consequences, prompting the international community to focus on nuclear disarmament and nonproliferation. International agreements like the Treaty on the Non-Proliferation of Nuclear Weapons aim to restrict the spread of nuclear arms and advance global disarmament efforts to uphold international peace and security.

6.1.4.2 *Military nuclear propulsion*

Nuclear power is primarily used in military applications for submarines and aircraft carriers, providing vessels with sustained propulsion. Nuclear reactors generate thermal energy through nuclear fission, which is then converted into electrical or mechanical energy to propel vessels. Nuclear-powered ships boast extended endurance, high speeds, and superior stealth capabilities. Nuclear submarines, in particular, can remain submerged for extended periods without surfacing for refueling, significantly enhancing naval strategic deterrence and operational flexibility.

6.1.4.3 *Military reconnaissance and protection*

Another critical military application of nuclear technology is radiation reconnaissance and protection. Military forces deploy various radiation detection devices to identify enemy nuclear weapon use or nuclear accidents, enabling timely countermeasures. In addition, nuclear protection equipment and technologies—such as radiation suits, shielding materials, and decontamination techniques—are vital for safeguarding personnel from radiation exposure. Command and control systems for nuclear weapons are essential for ensuring arsenal security and strategic stability.

6.1.4.4 *Evaluation*

The military application of nuclear technology demonstrates its formidable power and potential for devastation. Although nuclear weapons and nuclear-powered vessels provide strategic advantages to certain nations, they also entail significant risks and responsibilities that cannot be ignored. The international community continues to strive, through dialogue, cooperation, and legal frameworks, to reduce nuclear threats while safeguarding national security. Thus the military application of nuclear technology is not merely a technical issue but a complex matter involving international politics, diplomacy, and law.

6.2 The dual carbon goals and their demand for nuclear technology

6.2.1 Dual carbon goals and their value

6.2.1.1 Dual carbon goals and content

6.2.1.1.1 Dual carbon goals

The "dual carbon" goals—namely, "carbon peaking" and "carbon neutrality"—represent crucial strategic initiatives for countries in addressing climate change and promoting green, low-carbon development. The United Nations actively promotes global action on climate change and has provided a framework and guidance for countries to address climate change through a series of documents, including the United Nations Framework Convention on Climate Change, the Kyoto Protocol, and the Paris Agreement. International carbon peaking targets use 2030 and 2050 as key milestones to construct a blueprint for global climate governance. China's proposal of the "dual carbon" goals—achieving carbon peaking by 2030 and carbon neutrality by 2060—stems from multiple contexts and motivations, including the international climate change landscape, domestic sustainable development needs, economic transformation and upgrading, and the driving force of scientific and technological innovation. International financial institutions such as the World Bank and the Asian Development Bank have successively ceased financing coal-related projects to promote the development of global green finance. However, beginning in 2025, the Trump administration in the United States resumed the development of fossil fuels, and Europe also began to waver, casting a shadow over the realization of the "dual carbon" goals.

6.2.1.1.2 Carbon peak

By 2030 all nations must drive carbon dioxide emissions to peak and then decline. Developing countries must accelerate energy transitions and industrial upgrades, whereas developed nations must intensify emission reductions. China aims for its carbon dioxide emissions to peak around 2030, cease growth, and gradually decline thereafter.

6.2.1.1.3 Carbon neutrality

The international carbon neutrality goal seeks to balance global anthropogenic greenhouse gas emissions with natural absorption capacity to address the climate crisis. Under the Paris Agreement framework the world must achieve carbon neutrality around mid-century. To this end, countries will

Deeply transform energy structures to significantly increase renewable energy shares;

Advance low-carbon technological innovation in industries such as manufacturing and transportation; and

Strengthen forest conservation and ecological restoration to enhance natural carbon sinks.

By 2050 global carbon neutrality must be realized, balancing anthropogenic emissions with natural absorption. Through sustained efforts the global average temperature rise will be kept below 2°C above preindustrial levels, with efforts to limit it to 1.5°C, actively promoting the world's transition to a green, low-carbon development model. China's carbon neutrality goal is to achieve net-zero carbon dioxide emissions by 2060, meaning emissions and carbon sinks will offset each other through emission reductions and increased carbon sequestration.

6.2.1.2 *The demand for "dual carbon"*

6.2.1.2.1 The urgency of global climate change

Climate change stands as one of humanity's greatest challenges in the 21st century. The climate shifts driven by global greenhouse gas emissions have already triggered severe impacts, including increased extreme weather events, rising sea levels, and declining biodiversity. Addressing climate change demands concerted global action to reduce greenhouse gas emissions. In 2015 the Paris Agreement was adopted, with 195 countries committing to collectively limit global temperature rise to well below 2°C above preindustrial levels and pursue efforts to limit it to 1.5°C. As signatories, nations must fulfill their pledges and take concrete actions to reduce greenhouse gas emissions.

6.2.1.2.2 Environmental development needs

Rapid economic development has placed severe environmental and resource pressures on countries, including China. Environmental challenges, such as air pollution, rising temperatures, water scarcity, and land degradation, are increasingly prominent, urgently requiring a shift toward green, low-carbon, and sustainable development pathways. Reducing carbon dioxide emissions also helps decrease other pollutants such as sulfur dioxide, nitrogen oxides, and particulate matter. This improves air quality, protects ecosystems, and enhances public health. Energy conservation, emission reduction, and enhanced energy efficiency help reduce dependence on fossil fuels, promote resource conservation and recycling, and improve resource utilization efficiency.

6.2.1.2.3 Requirements for economic transformation and upgrading

Many economies are undergoing critical phases of transformation and upgrading, shifting from traditional, high-energy-consumption, high-emission, extensive growth models to green, low-carbon, efficient, and intensive development models. This involves fostering strategic emerging industries such as new energy, new materials, new energy vehicles, and energy conservation and environmental protection to create new economic growth drivers, enhance economic competitiveness, and strengthen resilience against risks. Accelerating the green transformation of traditional industries, promoting energy conservation and carbon reduction in sectors such as manufacturing, construction, and transportation, and elevating the green competitiveness of industrial chains are also essential.

6.2.1.2.4 Driving force of technological innovation

The development of low-carbon and green technologies provides the technical foundation for achieving the "dual carbon" goals. Countries, including China, have made significant progress in new energy technologies, smart grids, energy storage, and carbon capture and storage (CCS). Technological innovation can improve energy efficiency, reduce carbon intensity, and provide technical assurance for achieving carbon peak and carbon neutrality. Scientific and technological innovation drives the green transformation of energy, transportation, construction, and other sectors, promotes industrial restructuring and upgrading, and enhances the economy's sustainable development capacity.

6.2.1.3 Dual carbon goals value

6.2.1.3.1 Environmental significance

Carbon dioxide is the primary component of greenhouse gases. Reducing its emissions can effectively curb global warming and mitigate the impacts of climate change. Improving air quality and reducing fossil fuel use not only cuts carbon dioxide emissions but also reduces other pollutants (such as PM2.5, sulfur dioxide, and nitrogen oxides), thereby enhancing air quality and protecting the ecological environment.

6.2.1.3.2 Economic significance

Driving the transformation of traditional high-carbon industries toward low-carbon, green, and efficient models accelerates the development of emerging sectors such as clean energy, green manufacturing, and new energy vehicles, thereby optimizing and upgrading the economic structure. The dual carbon goals spur technological innovation in energy, environmental protection, transportation, and related fields, promoting R&D and application of relevant technologies. This, in turn, fuels the growth of high-tech industries and enhances national technological competitiveness [21,22].

6.2.1.3.3 Social significance

Enhancing public health: Reducing fossil fuel use and air pollution helps mitigate health risks and improve residents' quality of life. Creating employment opportunities: Developing green industries and clean energy sectors will generate substantial new jobs, boosting social employment and economic prosperity.

6.2.1.3.4 International significance

The dual carbon goals contribute to global climate governance. As the world's largest developing nation, China's commitment to these targets demonstrates its resolve and responsibility in addressing climate change, making a positive contribution to global climate governance [23]. Active participation in international climate negotiations and cooperation, strengthened collaboration with other nations in low-carbon technologies and green finance, enhances China's voice and influence in global environmental governance. This approach increases opportunities for international cooperation and elevates China's standing on the global stage [4].

6.2.2 Applications of nuclear technology in the energy sector

6.2.2.1 Nuclear fission power generation

6.2.2.1.1 Application characteristics

Nuclear power generation harnesses thermal energy from nuclear fission reactions to produce electricity, representing one of the primary applications of nuclear technology. Nuclear power plants control the heat generated by fission reactions to heat water, producing steam that drives turbines to generate electricity [24]. Its advantages include low carbon emissions, high energy density, and the provision of stable baseload power. Nuclear power plants emit virtually no carbon dioxide during operation, making them a clean energy source that helps reduce greenhouse gas emissions and mitigate global warming. Nuclear fuel possesses far greater energy density than fossil fuels—the energy released by the fission of 1 g of U-235 is equivalent to that from burning 2.7 t of coal. Nuclear plants deliver stable electricity unaffected by weather or seasonal variations, serving as a reliable power source.

6.2.2.1.2 Application in Europe and the United States

Nuclear fission power generation holds significant global prominence in Europe and the United States. The United States operates numerous nuclear power units, positioning it as a major nuclear power producer. It actively advances third- and fourth-generation advanced fission technologies, with small modular reactors (SMRs) also being a key development focus. France exhibits high dependence on nuclear power, with reactors supplying the majority of its electricity. Its mature pressurized water reactor (PWR) technology is being applied to new plant construction while extending the operational lifespan of existing reactors. In addition, European nations like the United Kingdom are adjusting their nuclear policies based on national circumstances, advancing project development and technological research. Overall, Europe and the United States possess advanced technology and extensive experience in nuclear fission power generation, continuously exploring innovations to enhance nuclear power's safety, economic viability, and sustainability.

6.2.2.1.3 China's application

China is actively constructing its independently developed third-generation nuclear technologies, Hualong One and Hualong One Plus, while also operating the world's most widely deployed reactor type, PWRs, such as those at Daya Bay and Qinshan nuclear power plants.

6.2.2.2 Nuclear fusion technology

6.2.2.2.1 Application characteristics

Nuclear fusion is considered one of the most promising clean energy technologies for the future. It generates vast amounts of energy through the fusion reaction of hydrogen isotopes (such as deuterium and tritium), similar to the reactions occurring inside the sun. The primary advantages of nuclear fusion lie in its zero carbon emissions and high safety profile. The fuel for nuclear fusion reactions is hydrogen isotopes, with helium being the main byproduct—no greenhouse gases are produced. Furthermore, nuclear fusion does not generate long-term radioactive waste. Reaction conditions are strictly controlled, and any loss of control automatically terminates the reaction, ensuring exceptional safety.

6.2.2.2.2 Applications in Europe and the United States

The US National Ignition Facility (NIF) achieved fusion gain $Q = 4$ in 2025, whereas Princeton's SMART device at the Plasma Physics Laboratory produced its first plasma beam. Europe's Joint Experimental Torus (JET) set an energy record of 69 MJ in 2023. France's WEST device successfully sustained plasma for 1337 seconds at temperatures exceeding 50 million °C. In October 2025 NVIDIA revealed it had developed an AI-driven digital twin fusion reactor. The October 2025 US–Japan Technology Prosperity Agreement specifically highlighted fusion energy as a priority area.

6.2.2.2.3 China's applications

China's independently developed nuclear fusion experimental device, "China Circulation Device II," has made multiple breakthroughs in this field. The subsequent inheritance and development include the "China Circulation 3" and even the future "China Circulation 4," which is a controllable nuclear fusion large-scale scientific device independently designed and developed by China. China Circulation III, also known as the new generation of "artificial sun," was equipped with an advanced magnetic field structure on June 13, 2024, in Chengdu, Sichuan. This breakthrough is of great significance for improving the control and operation capability of nuclear fusion devices, which can more accurately constrain and control high-temperature plasma and provide key technical support for the long-term stable operation of nuclear fusion devices. In January 2025 the EAST device in Hefei, China, achieved a high-confinement-mode plasma operation at 100 million °C for 1000 seconds, setting a new world record. In March 2025, for the first time, the temperature of both the atomic nucleus and electrons exceeded 100 million °C, marking significant progress in China's controllable nuclear fusion technology. In addition, Academician Zhang Jie pointed out that the fusion "dual cone ignition scheme" proposed by China has higher energy density and efficiency than the existing scheme in the United States.

6.2.2.3 Small modular reactor (SMR)

6.2.2.3.1 Application features

SMR is a new type of nuclear reactor characterized by low power, modular design, short construction period, and high safety. SMR can be flexibly deployed in remote areas, islands, and industrial parks to provide stable, clean electricity. Because of its lower power, SMR's safety design is more comprehensive and has higher inherent safety. SMR typically has a power of less than 300 MW and features modular design, short construction cycles, and relatively low costs.

6.2.2.3.2 European and American applications

SMRs in Europe and America are currently a hot research and application topic in the field of nuclear energy, suitable for power supply in remote areas, industrial energy use, and other scenarios. Several companies in the United States, such as NuScale, X-energy, and TerraPower, actively invest in research and development. Among them, NuScale is the first SMR company to obtain design certification from the US Nuclear Regulatory Commission (NRC), and technology giants such as Google and Amazon have also participated in investment cooperation [5]. The modular small nuclear reactor developed by NuScale has obtained design certification from the US NRC. In Europe the Eagle Alliance is advancing the development of the EAGLES-300 next-generation lead-cooled SMR, with plans to conduct its first demonstration before 2035 and achieve global commercialization by 2039. However, the development of SMR in Europe and America also faces challenges in cost control, supply chain bottlenecks, and regulation.

6.2.2.3.3 Chinese application

The small-scale PWR ACP100 developed by China National Nuclear Corporation (CNNC) is planned to be demonstrated and applied at the Changjiang nuclear power plant in Hainan. It is currently under active construction and has successfully achieved cold testing by October 2025.

6.2.2.4 Nuclear heating and cooling

6.2.2.4.1 Application features

Nuclear heating technology utilizes thermal energy from reactors for district heating, industrial heating, and cooling. During winter, it provides centralized heating for cities and communities, reducing reliance on coal-fired boilers and lowering carbon emissions. In industrial sectors, it supplies high-temperature heat sources for processes in petrochemicals, metallurgy, and other industries, enhancing energy efficiency and reducing carbon footprints.

6.2.2.4.2 Applications in Europe and America

Nuclear heating initiatives began earlier in Europe and America. For example, in 1956, the Calder Hall nuclear power plant in the United Kingdom supplied heat to nearby factories. In 1963 Sweden's Agesta nuclear power plant became the world's first reactor for civilian nuclear heating. Subsequently, countries such as Switzerland, France, and the United States also developed related applications. For instance, Switzerland's Beznau nuclear power plant has provided community heating through combined heat and power generation since 1980. Nuclear refrigeration: Europe and the United States have conducted research in this field. Nuclear adiabatic demagnetization refrigeration technology can cool macroscopic objects to temperatures around 10 μK, providing a crucial tool for extreme low-temperature physics research. However, nuclear refrigeration has seen relatively limited application in civilian sectors to date.

6.2.2.4.3 China's applications

China is also actively advancing nuclear heating initiatives. Examples include the Qinshan nuclear power plant's heating project, which utilizes residual heat from the plant to provide centralized heating for nearby cities, and the high-temperature gas-cooled reactor (HTGR), capable of supplying high-temperature industrial heat sources suitable for hydrogen production, seawater desalination, and other fields.

6.2.2.5 Nuclear-powered seawater desalination

6.2.2.5.1 Application features

Nuclear-powered seawater desalination utilizes thermal and electrical energy from nuclear power to desalinate seawater, producing freshwater resources to address water scarcity issues [25]. The desalination process demands substantial energy, and nuclear power provides a stable and efficient energy solution. This reduces reliance on fossil fuels while lowering carbon emissions and environmental pollution.

6.2.2.5.2 Application examples

Specific examples include the following: The Karachi nuclear power plant in Pakistan utilizes the waste heat from the plant to desalinate seawater and provide freshwater resources for the local area.

6.2.2.6 *Nuclear hydrogen production*

6.2.2.6.1 Application features

Nuclear hydrogen production is a clean way of producing hydrogen by electrolyzing water with nuclear energy. Hydrogen is an important component of future clean energy and can be used for fuel cells, industrial fuels, and more. The process of nuclear hydrogen production produces almost no carbon dioxide, and the hydrogen provided can be applied in various fields such as transportation, industry, and energy storage, promoting the development of hydrogen economy.

6.2.2.6.2 Application examples

A specific example involves utilizing high-temperature thermal energy from gas-cooled reactors for hydrothermal decomposition hydrogen production.

6.2.2.7 *Nuclear fusion and AI computing power*

6.2.2.7.1 Application characteristics

Nuclear fusion and AI computing power each possess distinct advantages and are interrelated in their application characteristics. When combined, AI accelerates nuclear fusion research by enabling plasma control, break prediction, experimental data mining, and more. Conversely, large-scale application of nuclear fusion could provide abundant energy for AI development, driving its further advancement. AI computing power is playing a pivotal role in fusion research. Fusion reactions demand precise control of high-temperature plasma—a complex, uncertain process challenging to master with traditional methods. AI enables rapid analysis of massive experimental datasets, uncovering underlying patterns to deepen scientists' understanding of plasma behavior. Conversely, fusion could potentially power AI in the future.

6.2.2.7.2 Application examples

In nuclear fusion research, AI computing power can simulate complex plasma behavior, optimize reactor design and control, and accelerate experimental data analysis. Should nuclear fusion achieve commercialization, it could provide AI with clean, stable, and massive energy support, creating a synergistic relationship between the two technologies.

6.2.3 The role and value of nuclear technology in achieving the dual carbon goals

6.2.3.1 *Objectives*

6.2.3.1.1 Provide stable low-carbon energy

Nuclear power generation is a quintessential example of low-carbon energy. During nuclear power plant operations, virtually no carbon dioxide is emitted, making it a crucial pathway for reducing greenhouse gas emissions. Compared to renewable energy sources such as wind and solar power, nuclear power provides stable baseload electricity unaffected by weather or seasonal variations. Therefore nuclear power is an essential component in achieving energy structure transformation, contributing to enhanced reliability and stability of the power system [6].

6.2.3.1.2 Driving energy structure transformation

The widespread adoption of nuclear energy helps optimize the energy mix and reduce dependence on fossil fuels. By developing nuclear power the proportion of coal, oil, and natural gas in energy consumption can be gradually reduced, thereby lowering carbon dioxide emissions. Furthermore the application of new nuclear technologies such as SMRs, molten salt reactors, and HTGRs enables clean energy provision across broader scenarios—including remote areas, islands, and industrial parks—facilitating distributed energy supply and regional energy self-sufficiency.

6.2.3.1.3 Providing industrial heat and hydrogen production

Nuclear heating technologies can supply high-temperature heat sources for industrial processes, replacing traditional coal-fired and gas-fired boilers to significantly reduce carbon emissions. In energy-intensive sectors such as petrochemicals and metallurgy, nuclear heating enhances energy efficiency and lowers carbon footprints. Nuclear energy can also be utilized for hydrogen production, employing high-temperature thermal energy from gas-cooled reactors to generate hydrogen through hydrothermal decomposition. The resulting hydrogen can be applied in fuel cells, industrial fuels, and other sectors, advancing the hydrogen economy and offering new pathways toward carbon neutrality.

6.2.3.1.4 Supporting desalination and district heating

Nuclear desalination technology harnesses thermal and electrical energy from nuclear power to convert seawater into freshwater, addressing water scarcity issues. Simultaneously, it reduces reliance on fossil fuels and lowers carbon emissions. Nuclear heating and cooling technologies can provide centralized heating for cities and communities, particularly during winter, reducing the use of coal-fired boilers and further decreasing carbon emissions. These applications not only enhance quality of life but also play a significant role in reducing carbon emissions.

6.2.3.2 Value

6.2.3.2.1 Promoting technological innovation

Developing nuclear technology and its applications can drive the growth of related high-tech industries, fostering technological innovation and economic transformation. The advancement of nuclear technology requires substantial investment in scientific research and the cultivation of technical talent, which supports the development of new productive forces—including AI—and enhances national scientific strength and international competitiveness. Simultaneously, nuclear technology development propels technological innovation, laying a solid foundation for achieving a green and sustainable future.

6.2.3.2.2 Economic development

Nuclear technology plays a pivotal role in achieving the dual carbon goals. By providing stable low-carbon energy, driving energy structure transformation, supplying industrial heat and hydrogen production, supporting desalination, and enabling district heating, nuclear technology offers robust support for reducing carbon emissions and achieving carbon neutrality. Furthermore the construction and operation of nuclear energy projects stimulate the growth of related industrial chains, create substantial employment opportunities, and propel sustainable economic development [7].

6.3 Demand for nuclear technology amid the pandemic

6.3.1 Demand for nuclear technology in pandemic healthcare

6.3.1.1 The socioeconomic impacts and demands of the COVID-19 pandemic

6.3.1.1.1 Pandemic impacts

The 2003 SARS outbreak and the 2020 COVID-19 pandemic delivered massive shocks to the global economy. Lockdown measures and restrictions on movement led to a significant reduction in economic activity, severely impacting numerous industries—particularly service sectors such as tourism, aviation, catering, and retail. Globally, rising unemployment, supply chain disruptions, and economic recessions became widespread challenges. Concurrently the pandemic profoundly affected mental health, with prolonged isolation and social distancing significantly increasing psychological issues such as loneliness, anxiety, and depression.

6.3.1.1.2 Application needs

At this stage, mental health support and intervention measures have become particularly crucial. The global spread of COVID-19 serves as a stark reminder that human society must respond swiftly and collaborate closely when confronting sudden public health emergencies. Only through the application of scientific prevention and control measures—including nuclear technology—alongside public participation and international cooperation can we ultimately overcome this threat and restore normal social and economic order.

6.3.1.2 Transmission hazards and control measures of COVID-19

6.3.1.2.1 Transmission hazards of COVID-19

COVID-19 is an acute respiratory infectious disease caused by the novel coronavirus (SARS-CoV-2). Since its initial reporting in Wuhan, Hubei Province, China, in December 2019, the disease has rapidly spread globally, triggering an unprecedented pandemic with profound impacts on human health, socioeconomic systems, and daily life. The novel coronavirus belongs to the coronavirus family, named for the spike proteins protruding from its surface, which appear crown-like under a microscope. SARS-CoV-2 spreads rapidly among people through respiratory droplets, contact transmission, and aerosol transmission. The incubation period typically ranges from 2 to 14 days, with most cases developing symptoms within 5–6 days after infection. COVID-19 symptoms vary widely, ranging from asymptomatic carriers to severe pneumonia. Primary symptoms include fever, dry cough, fatigue, muscle aches, sore throat, loss of taste and smell, and shortness of breath. Some patients may experience gastrointestinal symptoms such as diarrhea and vomiting.

6.3.1.2.2 Nuclear technology prevention measures

Nuclear technology has emerged as a vital tool in combating the pandemic. In testing, radionuclide labeling technology has been applied to develop nucleic acid detection reagents, significantly enhancing test sensitivity and accuracy. This enables rapid screening of infected individuals, buying critical time for epidemic control. In addition, nuclear technology supports vaccine development through irradiation techniques, accelerating the process while ensuring vaccine safety and efficacy—a powerful boost to the global fight against the pandemic. Leveraging its unique technological advantages, nuclear technology has played a pivotal role across multiple stages of COVID-19 prevention and control, providing robust support for curbing the spread of the virus and safeguarding public health.

6.3.1.3 From diagnostic methods to vaccination and treatment measures

6.3.1.3.1 Diagnostic methods and vaccination

The primary method for diagnosing COVID-19 involves detecting viral nucleic acids in nasopharyngeal or oropharyngeal swabs using reverse transcription polymerase chain reaction (RT-PCR). Rapid antigen tests and antibody tests are also commonly used diagnostic tools. Rapid antigen tests identify current infections, whereas antibody tests detect past infections. Vaccines play a crucial role in controlling the COVID-19 pandemic. Since 2020 multiple COVID-19 vaccines have received emergency use authorization and been deployed globally. These vaccines have significantly contributed to preventing infections, reducing severe illness rates, and lowering mortality rates.

6.3.1.3.2 Therapeutic measures including nuclear technologies

For treatment, mild cases typically receive symptomatic management such as fever reduction and cough suppression. Severe cases may require supportive measures such as oxygen therapy or mechanical ventilation. Certain antiviral drugs, such as remdesivir, and immunomodulators like dexamethasone have demonstrated efficacy in treatment. In therapeutic applications, radioisotope drugs show potential for treating severe COVID-19 patients. For example, leveraging the targeted therapeutic properties of radionuclides can reduce pulmonary inflammatory responses and aid patient recovery. In disinfection applications, irradiation technology utilizes ionizing radiation to sterilize medical devices and protective gear. Compared to traditional methods, it offers advantages of high efficiency, residue-free operation, and strong penetration capabilities. This technology rapidly eliminates viruses without compromising material integrity, ensuring the safe supply of medical resources.

6.3.2 Radiation sterilization and its nuclear technology requirements

6.3.2.1 Definition of irradiation sterilization

Irradiation sterilization technology utilizes high-energy electromagnetic radiation (such as gamma rays or X-rays) or high-energy particles (such as electron beams) to inhibit or kill microorganisms.

6.3.2.2 Classification of radiation sources

Primary radiation sources include gamma rays, electron beams, and X-rays.

6.3.2.3 Mechanism of action

The mechanism of nuclear irradiation sterilization involves using high-energy radiation to damage microorganisms' DNA, RNA, and other vital molecular structures, rendering them incapable of reproduction or causing direct death.

6.3.2.4 Advantages of irradiation technology

6.3.2.4.1 Efficient sterilization in short timeframes

Nuclear irradiation technology achieves highly effective sterilization within brief periods, making it suitable for large-scale disinfection of medical devices and supplies to meet urgent demands during epidemics.

6.3.2.4.2 Avoids chemical contamination and potential toxicity risks

Unlike traditional chemical disinfection methods, nuclear irradiation does not use chemical agents, leaving no chemical residues and eliminating risks of chemical contamination and potential toxicity.

6.3.2.4.3 Preserves quality and functionality

Conducted at ambient temperatures, nuclear irradiation is suitable for temperature-sensitive medical devices and pharmaceuticals without compromising their quality or functionality.

6.3.2.4.4 Ensures complete sterility

Gamma rays and X-rays in nuclear irradiation possess strong penetrating power, enabling comprehensive sterilization of items both inside and outside packaging to guarantee thorough sterility.

6.3.2.5 Application scope

6.3.2.5.1 Scope

It is widely applied across food, medical devices, pharmaceuticals, and other sectors to ensure product hygiene and safety while extending shelf life. Nuclear irradiation technology effectively eliminates various pathogens, enhancing product safety and quality.

6.3.2.5.2 Applications

6.3.2.5.2.1 Sterilization of single-use medical devices Common disposable items, such as syringes, surgical instruments, gauze, and dressings, undergo irradiation to guarantee sterility during use, preventing infection risks from microbial contamination.

6.3.2.5.2.2 Disinfection of protective gear During the COVID-19 pandemic, demand for protective equipment such as masks, coveralls, and face shields for healthcare workers surged. Nuclear irradiation technology was employed to disinfect these items, ensuring their safe use. In some cases, it was also used to sterilize reusable protective gear, alleviating supply shortages.

6.3.2.5.2.3 Sterilization of pharmaceuticals and medical supplies Microbial content must be strictly controlled during pharmaceutical production, particularly for injectable drugs. Nuclear irradiation effectively inactivates microorganisms in medications, ensuring sterility during storage and use. Medical supplies, such as dressings, catheters, and surgical sutures, achieve sterility through irradiation processing.

6.3.2.5.2.4 Environmental and waste treatment Nuclear irradiation technology can be applied for medical environment disinfection, particularly in critical areas such as intensive care units and operating rooms, ensuring complete elimination of pathogenic microorganisms. Medical waste may contain high concentrations of viruses requiring safe disposal. Nuclear irradiation provides efficient sterilization, reducing environmental pollution and the risk of secondary transmission.

6.3.3 Demand for nuclear technology in pharmaceuticals

6.3.3.1 *Definition of radionuclide drugs*

Nuclide drugs, also known as radiopharmaceuticals, refer to drugs containing radioactive isotopes, mainly used for medical imaging diagnosis and treatment.

6.3.3.2 *Classification of nuclear drugs*

Common diagnostic radioactive isotopes include technetium-99, iodine-123, and fluorine-18. Common therapeutic radioactive isotopes include iodine-131, strontium-89, and radium-223.

6.3.3.3 *Mechanism of action*

Nuclide drugs release radiation or particles through the decay of radioactive isotopes, which are used for diagnosing or treating diseases. According to their different uses, radiopharmaceuticals can be divided into diagnostic radiopharmaceuticals and therapeutic radiopharmaceuticals. Diagnostic radiopharmaceuticals guide radioactive isotopes to specific parts of the body through specific biomolecules such as sugars, proteins, and antibodies. These radioactive nuclides are captured by medical imaging equipment [such as PET and single photon emission computed tomography (SPECT)] by emitting gamma rays or positrons, thereby displaying functional and structural information in the body on images. Therapeutic radiopharmaceuticals release high-energy particles (such as alpha and beta particles) to locally irradiate diseased tissues, damaging the DNA of cancer cells and killing them or inhibiting their growth.

6.3.3.4 *Advantages of radionuclide drugs*

6.3.3.4.1 Improve the accuracy of diagnosis and treatment

Nuclide drugs can selectively concentrate in specific tissues or organs, improving the accuracy of diagnosis and treatment.

6.3.3.4.2 Directly acting on the affected area

Diagnostic radionuclide drugs obtain in vivo information through in vitro imaging technology, reducing patient pain and risk, and can directly act on the lesion site.

6.3.3.4.3 Enhanced therapeutic outcomes

They minimize damage to surrounding healthy tissues, thereby improving treatment efficacy.

6.3.3.5 Application scope

Radionuclide drugs find extensive use in cancer diagnosis and treatment, cardiovascular disease diagnosis, and thyroid disease diagnosis and management.

6.3.4 Medical imaging's demand for nuclear technology

6.3.4.1 Definition of nuclear medicine imaging technology

Nuclear medicine imaging technology is a medical specialty that utilizes radiopharmaceuticals (radioisotope drugs) for diagnosis and treatment. It integrates nuclear physics and medical imaging by employing compounds labeled with radioactive isotopes to track their distribution and metabolism within the body, thereby obtaining functional and structural information about specific organs or tissues.

6.3.4.2 Types of nuclear medicine imaging technology

Nuclear medicine imaging primarily includes SPECT, PET, boron neutron capture therapy (BNCT) for cancer treatment, PET–Computed Tomography (PET–CT) imaging, next-generation MI techniques, and Magnetic Resonance Imaging (MRI). Details are provided in Section 3.4.1.

6.3.4.3 Imaging mechanism

The fundamental principle of nuclear medicine imaging involves introducing radiopharmaceuticals into the body. These drugs circulate via the bloodstream to reach specific organs or tissues. As the radiopharmaceuticals decay within the body, they emit radiation (such as gamma rays or positrons). This radiation is captured by external detection devices (e.g., gamma cameras or PET scanners). Through computer processing, functional images of specific anatomical regions can be generated.

6.3.4.4 Advantages of nuclear medicine imaging technology

6.3.4.4.1 Aids diagnosis, monitoring, and research

Nuclear medicine imaging plays a vital role in early disease diagnosis, treatment monitoring, and research.

6.3.4.4.2 Facilitates early diagnosis and treatment

These images provide information on organ function, blood flow, metabolic activity, and more, supporting early diagnosis and treatment [8].

6.3.4.5 Application scope

6.3.4.5.1 Scope

Widely applied clinically in tumor diagnosis and monitoring, cardiac disease diagnosis, brain disorder research, bone disease assessment, and thyroid function evaluation.

6.3.4.5.2 Applications

6.3.4.5.2.1 SPECT applications SPECT is suitable for diagnosing and treating cerebrovascular diseases, bone disorders, and other functional and structural conditions.

6.3.4.5.2.2 PET applications PET is primarily used for early cancer diagnosis and monitoring, cardiac function assessment, and brain function research.

6.3.4.5.2.3 MRI applications It is used in medical diagnosis to precisely visualize tumors, inflammation, injuries, and other pathologies in the brain, spinal cord, joints, etc. It is also employed for functional imaging of brain activity, myocardial perfusion, and biological tissue research.

6.3.4.5.2.4 Combined applications PET/CT and SPECT/CT integrate functional imaging from PET or SPECT with anatomical imaging from CT. This combination enables precise tumor localization, treatment efficacy assessment, and radiation therapy planning guidance, significantly enhancing diagnostic accuracy and precision.

6.3.5 Radiation therapy for tumors: Requirements for nuclear technology

6.3.5.1 Definition of radiation therapy technology

Radiation therapy for tumors employs high-energy radiation to inhibit cancer cell growth or destroy them. It can be used as a standalone treatment or combined with other therapies such as surgery and chemotherapy.

6.3.5.2 Types of radiation therapy

Primary types include external beam radiation therapy (EBRT), brachytherapy, radionuclide therapy, and proton/heavy ion therapy.

6.3.5.3 Mechanism of action

The fundamental principle of radiation therapy involves using high-energy rays—such as X-rays, gamma rays, electron beams, or proton beams—to damage cancer cell DNA, thereby preventing their growth and division.

6.3.5.4 Advantages of radiation therapy

Although radiation causes nonspecific damage to cells, cancer cells exhibit poorer repair capabilities, making radiation therapy more effective at destroying them. Normal cells also sustain some damage, but their robust repair mechanisms allow gradual recovery during treatment.

6.3.5.5 Scope of application

6.3.5.5.1 Scope

Radiation therapy is applied to solid tumors at all stages and certain hematologic malignancies. It can serve as a standalone treatment for early-stage tumors, complement surgery and chemotherapy in comprehensive management of advanced-stage cancers, and alleviate symptoms like pain in late-stage tumors to enhance patient quality of life.

6.3.5.5.2 Applications

6.3.5.5.2.1 External beam radiation therapy EBRT is the most common form of radiotherapy. It uses external equipment to generate high-energy radiation beams focused on the tumor area, suitable for various solid tumors such as head and neck cancers, lung cancer, breast cancer, and prostate cancer.

6.3.5.5.2.2 Brachytherapy Brachytherapy involves implanting radioactive material directly into or near the tumor site for localized irradiation. It is commonly used to treat cervical cancer, prostate cancer, breast cancer, and skin cancer.

6.3.5.5.2.3 Radionuclide therapy applications Radionuclide therapy utilizes radioactive isotopes administered intravenously or orally. These isotopes selectively accumulate in tumors or metastatic sites for treatment, primarily used for thyroid cancer and bone metastases.

6.3.5.5.2.4 Proton and heavy ion therapy applications Proton and heavy ion therapy employs precisely directed beams of protons or heavy ions to irradiate tumors while minimizing damage to surrounding healthy tissues. It is suitable for refractory tumors and those unresponsive to conventional radiotherapy, such as chordomas and chondrosarcomas.

6.4 Reliability in nuclear technology engineering applications

6.4.1 Nuclear power plant reliability

6.4.1.1 Design optimization

6.4.1.1.1 Purpose

During the design process of a nuclear power plant, optimizing the reactor design is crucial for ensuring its stability and reliability.

6.4.1.1.2 Methods

6.4.1.1.2.1 Strictly adhere to standards and specifications The reactor design of a nuclear power plant must comply with stringent safety standards and specifications to ensure stability under both normal operation and accident conditions.

6.4.1.1.2.2 Utilize high-strength, corrosion-resistant materials The structural design must account for high temperatures, pressures, radiation, and other factors, employing high-strength, corrosion-resistant materials to prevent failures such as fractures or leaks during long-term operation [9,10].

6.4.1.1.2.3 Optimize fuel loading and thermodynamic design Fuel loading and thermodynamic design require rational planning to enhance fuel utilization and thermal efficiency, ensuring sustained stable reactor operation.

6.4.1.2 Process control

6.4.1.2.1 Purpose

Process control in nuclear power reactors is a critical way to ensure stability and reliability.

6.4.1.2.2 Methods

6.4.1.2.2.1 Implement real-time monitoring and control of parameters Parameters, such as reactor temperature, pressure, and flow rate, require real-time monitoring and control to ensure operation within safe limits.

6.4.1.2.2.2 Establish a comprehensive automated control system A comprehensive automated control system should be established to promptly respond to and address abnormal conditions, ensuring the reactor can self-regulate and operate stably [11–13].

6.4.1.2.2.3 Implementing rational fuel management and discharge control Rational fuel management and discharge control are also vital components for ensuring reactor stability. Through scientific management of fuel usage and processing the environmental and human health impacts of radioactive waste from fuel are minimized.

6.4.1.3 Safety assurance

6.4.1.3.1 Purpose

Safety assurance for nuclear power plant reactors is the foundation for guaranteeing their stability and reliability.

6.4.1.3.2 Methods

6.4.1.3.2.1 Establish comprehensive safety management systems and operating procedures Nuclear power plants must develop comprehensive safety management systems and operating procedures, clearly defining all operational workflows and emergency measures to ensure timely and effective responses under any circumstances [14].

6.4.1.3.2.2 Implement scientific radiation protection and environmental monitoring Conduct scientific radiation protection and environmental monitoring to promptly detect and address potential radiation leaks and environmental pollution issues, safeguarding the health and safety of personnel and surrounding residents.

6.4.1.3.2.3 Maintain open communication and transparency Nuclear power plants should also foster open communication and transparency with relevant authorities and the public, promptly and accurately conveying information about nuclear safety to enhance societal trust and support for nuclear facilities.

6.4.1.4 Implementation methods

6.4.1.4.1 Purpose

Ensuring reactor stability and reliability requires comprehensive consideration across multiple dimensions, including design optimization, process control, and safety assurance.

6.4.1.4.2 Approach

Only through scientific design, rational control, and effective safety safeguards can nuclear power plants achieve long-term stable operation, providing reliable and clean energy to society. As one of the world's most critical forms of power generation the stability and reliability of the reactor—the core equipment—are paramount to the stable operation of nuclear power plants. Three key areas—design optimization, process control, and safety safeguards—can be explored to ensure reactor stability and reliability.

6.4.2 Nuclear weapons reliability

6.4.2.1 Reliability

6.4.2.1.1 Definition

The probabilistic measure of dependability is termed "reliability." Reliability is one of the key indicators of the tactical and technical performance of nuclear weapons. Nuclear weapon reliability is closely linked to design, manufacturing, and other stages of the weapon system. Reliability design serves as the foundation.

6.4.2.1.2 Metrics

6.4.2.1.2.1 Types The primary quantitative reliability metrics are deployability rate and nuclear detonation reliability.

6.4.2.1.2.2 Deployability rate Deployability rate refers to the probability of timely deployment when required.

6.4.2.1.2.3 Nuclear detonation reliability Nuclear detonation reliability denotes the probability that a nuclear weapon will achieve detonation according to specified performance criteria under normal deployment conditions.

6.4.2.2 Reliability design

6.4.2.2.1 Types

Reliability design for nuclear weapons permeates both physical and engineering design.

6.4.2.2.2 Physical design

In physical design, meticulous selection of all physical parameters is required, considering various influencing factors and incorporating appropriate margins [15]. Comprehensive verification of the stability of nuclear device performance is achieved through detonation tests under various conditions and a sufficient number of nuclear explosion tests.

6.4.2.2.3 Engineering structural design

Engineering structural design must comprehensively account for environmental load factors, storage corrosion, aging, material compatibility, and other influences to incorporate margin design.

6.4.2.2.4 Approaches

Considerations include employing multiple fuses with differing mechanisms within the nuclear weapon detonation control system; implementing redundant component designs and derating measures for vulnerable points; and rationally determining component clearances and tolerances.

6.4.2.3 Reliability validation

6.4.2.3.1 Test validation

Under specified environmental conditions, analyze system and component responses in extreme environments. Conduct various reliability tests and flight tests for nonnuclear systems to validate product reliability. The advanced light source (ALS) at Lawrence Berkeley National Laboratory in the United States utilizes some of its experimental platforms to support nonexplosive testing research on nuclear materials for the National Nuclear Security Administration (NNSA).

6.4.2.3.2 Mathematical and statistical verification

Given the prohibitive cost of nuclear testing, it is impractical to assess nuclear device reliability through statistically significant large-scale nuclear tests. However, for nonnuclear systems—such as electronic components and explosives—reliability should be evaluated using statistically meaningful subsamples.

6.4.2.4 Factors affecting the reliability of stockpiled nuclear weapons

6.4.2.4.1 Primary factors

Nuclear weapons spend the vast majority of their service life in storage. Key factors affecting reliability during storage include material decay and aging, metal corrosion, and degradation of organic materials.

6.4.2.4.2 Accelerating factors

Contact between incompatible materials accelerates corrosion. Particularly the aging and corrosion of chemically and physically unstable radioactive nuclear materials (e.g., plutonium) directly impact nuclear performance.

6.4.2.5 Reliability measures

6.4.2.5.1 Design protection for nuclear materials

Design considerations must fully account for the storage stability and material compatibility of all components (especially nuclear materials), as well as their environmental adaptability. For instance, plutonium warheads employ sealed structures to prevent chemical reactions and corrosion from external exposure.

6.4.2.5.2 Nuclear stockpile inspection and maintenance

Enhance regular monitoring and maintenance during stockpile storage. Periodically, sample stored products for functional testing, such as detonator synchronization verification and detonation simulation tests, to assess storage performance. Components with limited lifespans (e.g., tritium-containing parts) should be replaced regularly.

6.4.2.5.3 Nuclear weapons reliability assurance

6.4.2.5.3.1 Nonnuclear system flight tests Conduct nonnuclear system flight tests to verify system reliability.

6.4.2.5.3.2 Sampling tests Prior to the Comprehensive Nuclear-Test-Ban Treaty, stockpile reliability could be verified through sampling tests of stockpiled nuclear weapons. After the treaty's implementation, ensuring the reliability of nuclear weapons stockpiles without nuclear testing capabilities has become a critical challenge for all nuclear-weapon states.

6.4.3 Nuclear medical reliability

6.4.3.1 Objectives of nuclear medicine

Nuclear medicine plays a vital role in clinical practice, providing physicians with more accurate and earlier diagnostic tools while offering patients more effective and safer treatment options. With continuous technological advancements the application prospects of nuclear medicine in the medical field will become increasingly broad.

6.4.3.2 Characteristics of nuclear medicine

As a specialized medical discipline, nuclear medicine features noninvasive diagnostics, cellular-level testing, multifunctionality, radiation safety, and early diagnosis capabilities.

6.4.3.3 Applications of nuclear medicine

Nuclear medicine primarily has two applications: First, nuclear medicine imaging, including SPECT and PET, enables the acquisition of molecular-level information about tissues or organs without causing trauma to the human body. It is a crucial means for the early diagnosis of major diseases such as cardiovascular and cerebrovascular diseases, neurodegenerative diseases, and cancer. Second, radiation therapy for tumors, primarily utilizes alpha, beta, and gamma rays from radioactive isotopes, along with X-rays, electron beams, proton beams, and other particle beams generated by X-ray machines or accelerators to treat cancerous and other pathological tissues.

6.4.3.4 Reliability assessment criteria

6.4.3.4.1 Noninvasive diagnosis

Nuclear medicine diagnostics require no surgery or incisions. Radioactive substances are introduced into the body via injection of radionuclides or ingestion of radiopharmaceuticals. Imaging or measurements are then performed using the emitted gamma or beta rays from these substances. This noninvasive approach reduces patient discomfort and recovery time while enhancing diagnostic accuracy.

6.4.3.4.2 Cellular-level testing

Nuclear medicine diagnostics enable observation of tissue and organ function and metabolic status at the cellular level. The unique properties of radionuclides allow them to enter the body in minute quantities and emit specific radiation within biological systems. Measurement and analysis of this radiation yield more precise diagnostic outcomes.

6.4.3.4.3 Multifunctionality

Nuclear medicine techniques serve diverse diagnostic and therapeutic applications. For instance, it diagnoses heart disease by injecting radionuclides to visualize cardiac blood flow; treats cancer by using radionuclide radiation to destroy tumor cells; and manages thyroid disorders through oral radioactive iodine therapy for thyroid cancer. This versatility makes nuclear medicine indispensable in medical practice.

6.4.3.4.4 Radiation safety

The radioactive isotopes used in nuclear medicine undergo strict control in both dosage and duration to ensure radiation exposure remains within safe limits for the human body. Commonly used isotopes in clinical practice have short half-lives and low radiation energy, allowing them to decay completely within the body in a short time, thereby minimizing radiation damage [16].

6.4.3.4.5 Early diagnosis

Nuclear medicine diagnostics enable physicians to detect diseases at their earliest stages. This holds significant value for the early diagnosis of certain conditions. For example, nuclear medicine techniques can detect tumors, skeletal disorders, and cardiovascular diseases at an early stage. This enhances treatment success rates and prolongs patient survival.

References

[1] A. Peakman, A. Detkina, B. Lindley, The role of nuclear fission technology in decarbonization, in: Energy Clim. Change, 2025, pp. 455–474.

[2] L.J. Dang, X.L. Li, W.R. Zhang, et al., Design and Implementation of DCS System MTS for nuclear power plant based on plug-in technology, in: P. Gu, Y. Xu, W. Chen, Z. Wang, Y. Sun, Z. Liu (Eds.), New Energy Power Generation Automation and Intelligent Technology. SICPNPP 2024. Lecture Notes in Electrical Engineering, 1249, Singapore, Springer, 2024, pp. 49–62, doi:10.1007/978-981-97-7051-9_4.

[3] L. Yongchao, L. Tong, X. Kai, et al., Research on sensor data optimization technology for thermal hydraulic experiment of nuclear reactor, Nucl. Eng. Des. (2024) 423, doi:10.1016/j.nucengdes.2024.113176.

[4] A. Huang, M. Guo, L. Dai, et al., Budgeting for a greener future: asymmetric nexus between nuclear energy technology budgets and CO_2 emissions, Technol. Forecast. Soc. Change 202 (2024), doi:10.1016/j.techfore.2024.123321.

[5] C. Granger, Sheffield Forgemasters in nuclear welding technology development, Mach. Mark. (TN.6330) (2024).

[6] Y. Guo, K. Huang, Z. Lyu, et al., Radiation safety management of scientific research projects of nuclear technology and application disciplines in colleges and universities, Exp. Technol. Manage. 40 (2) (2023) 1–6, doi:10.16791/j.cnki.sjg.2023.02.001.

[7] G. Currie, S. Robbie, P. Tually, ChatGPT and patient information in nuclear medicine: GPT-3.5 versus GPT-4, J. Nucl. Med. Technol. 51 (4) (2023) 307–313, doi:10.2967/jnmt.123.266151.

[8] A. Ketharam, D. Pletser, N. Evans, et al., Field-assisted processing of Ce–zirconolite ceramics for nuclear applications, Int. J. Appl. Ceram. Technol. (2025), doi:10.1111/ijac.15119.

[9] X.L. Zhu, C.L. Dong, Y. Li, et al., The access control technology of domestic gateway in nuclear safety-level DCS system, International Symposium on Software Reliability, Industrial Safety, Cyber Security and Physical Protection for Nuclear Power Plant, Springer, 2024, doi:10.1007/978-981-97-7051-9_46.

[10] D. Dong, Z. Wang, J. Guan, et al., Research on key technology and application progress of rescue robot in nuclear accident emergency situation, Nucl. Eng. Technol. 57 (6) (2025), doi:10.1016/j.net.2025.103457.

[11] S.G. Mashnik, A.J. Sierk, Proceedings of the International Conference on Nuclear Data for Science and Technology, Nucl. Data Sheets (2002) 118.

[12] Y. Li, M.X. Liu, H. Peng, et al., Research and application of endogenous security active defense technology for domestic nuclear safety-level gateway equipment, in: International Symposium On Software Reliability, Industrial Safety, Cyber Security and Physical Protection for Nuclear Power Plant, Singapore, Springer, 2023, doi:10.1007/978-981-99-3455-3_20.

[13] Y.Q. Chen, B.H. Yan, The technology of shielding design for nuclear reactor: a review, Prog. Nucl. Energy 161 (2023) 1–2, doi:10.1016/j.pnucene.2023.104741.

[14] A. Verma, Reckoning with the wicked problems of nuclear technology: pedagogical philosophy, design, and method underlying a course on nuclear technology, policy, and society, Front. Nucl. 3 (2024) 1382695.

[15] M. Xie, Y. Chen, L. Yu, et al., Technical research on rapid source term calculation and prediction technology of nuclear accidents based on Bayesian network, J. Nucl. Sci. Technol. 61 (8) (2024) 15, doi:10.1080/00223131.2023.2294193.

[16] M. Teshome, Ensuring energy security through expanded access to nuclear technology for peaceful uses and the challenge of a nuclear taboo, Eur. J. Law Political Sci. (2023), doi:10.24018/ejpolitics.2023.2.2.76.

[17] M. Trtica, J. Savovic, M. Kuzmanovic, D. Rankovic, J. Stasic, Libs hydrogen isotopes detection: significance in nuclear/fusion technology, J. Appl. Spectrosc. 90 (6) (2024) 1318–1324, doi:10.1007/s10812-024-01669-0.

[18] L.G.M. Siqueira, R.F.D. Assis, J. César Montecinos, et al., Implementation of a business intelligence system in the Brazilian nuclear industry: an action research, Procedia Comput. Sci. 232 (2024) 956–965, doi:10.1016/j.procs.2024.01.095.

[19] W. Zhizhan, Z. Shende, Z. Lifa, et al., Application of nuclear magnetic resonance logging technology in physical property analysis of rock, Pet. Geol. Exp. 27 (6) (2025), doi:10.1016/j.molcatb.2005.02.001.

[20] N. Otuka, O. Schwerer, S. Dunaeva, et al., The role of the nuclear reaction data centres in experimental nuclear data knowledge sharing, J. Korean Phys. Soc. 59 (2) (2011) 1292, doi:10.3938/jkps.59.1292.

[21] M. Balat, The role of nuclear power in global electricity generation, Energy Sources, B: Econ. Plan. Policy 2 (4) (2007) 381–390, doi:10.1080/15567240600705458.

[22] M.B. Abdelhamid, et al., The role of nuclear power in reducing risk of the fossil fuel prices and diversity of electricity generation in Tunisia: a portfolio approach, J. Fusion Energy (2010), doi:10.1007/s10894-009-9258-2.

[23] M.D.A.M.A. Ahmed, The role of nuclear energy in international relations, the Russian-Ukrainian war as a model, Int. J. Humanit. Educ. Res. (2023), doi:10.47832/2757-5403.18.28.

[24] A. Gour, R.K. Behera, T. Mathews, Novel approach for network data exchange among safety-related control systems of nuclear reactors, Nucl. Technol. 211 (2) (2025) 318–331, doi:10.1080/00295450.2024.2325749.
[25] Z.J. Ingersoll, et al., NuScale small modular reactor for Co-generation of electricity and water[J], Desalination 340 (1) (2014) 84–93, doi:10.1016/j.desal.2014.02.023.

Chapter 7

Nuclear technology and human development

Shang Mao [a], Tao Zhou [b], Chunhui Xue [b], Yao Yao [b], Wenbin Liu [b] and Jianyu Tang [b]

[a] *School of Energy and Power Engineering, Zhengzhou University of Light Industry, Zhengzhou, Henan, China,* [b] *Department of Nuclear Science and Technology, Institute of Nuclear Thermal-hydraulic Safety and Standardization, Southeast University, Nanjing, Jiangsu, China*

7.1 Nuclear technology and industrial development

7.1.1 Development of the nuclear industry supply chain

7.1.1.1 Key stages of the nuclear industry

7.1.1.1.1 Nuclear fuel mining and processing

The upstream stage involves uranium ore exploration, mining, and refining. Uranium mining requires specialized equipment and technology, driving the development of mining and metallurgical industries. Uranium refining and processing necessitate chemical engineering techniques, imposing high-purity and high-quality demands on the chemical industry.

7.1.1.1.2 Nuclear power plant construction and operation

The midstream segment includes the design, construction, and daily operation of nuclear power plants. These activities require extensive engineering construction and equipment manufacturing—such as reactors, steam generators, and cooling systems—involving multiple industries, including mechanical manufacturing, electronics, and electrical engineering [1].

7.1.1.1.3 Nuclear waste disposal and facility decommissioning

Downstream processes involve the safe disposal of nuclear waste and the decommissioning of nuclear facilities. These tasks demand advanced chemical processing and materials science technologies, along with specialized engineering management and construction capabilities [2], driving the development of new materials and environmental protection industries.

Nuclear Science, Technology, and Civilization. DOI: https://doi.org/10.1016/B978-0-443-13941-3.00007-0
Copyright © 2026 Elsevier Inc. All rights are reserved, including those for text and data mining, AI training, and similar technologies.

7.1.1.2 The nuclear industry's catalytic role in related sectors

7.1.1.2.1 Driving development across industrial sectors

Every segment of the nuclear industry supply chain requires precision technology and equipment, thereby stimulating growth in multiple industrial sectors.

7.1.1.2.2 Advancement of the metallurgical industry

Nuclear industry growth imposes new demands on materials science, chemical engineering processes, and engineering technologies. For instance, constructing nuclear power plants requires high-strength, corrosion-resistant specialty steels, driving technological progress in metallurgy.

7.1.1.2.3 Advancement of the chemical industry

Nuclear fuel production necessitates high-purity chemicals, propelling upgrades within the chemical sector. This not only creates demand for high-tech positions but also stimulates growth in supporting industries, providing economic growth momentum for local regions.

7.1.1.2.4 Driving the development of construction and transportation industries

Nuclear power plant construction demands substantial quantities of building materials, mechanical equipment, and transportation services.

7.1.1.2.5 Driving regional economic development

The production and service demand across these industries stimulate every link in the upstream and downstream industrial chains, driving comprehensive regional economic development.

7.1.1.3 Nuclear industry development in different countries

7.1.1.3.1 United States

The US nuclear industry experienced rapid growth in the mid-20th century, driving advancements in nuclear engineering and materials science while spurring prosperity in related industries such as mechanical manufacturing, electronics, and electrical engineering. For example, Oak Ridge National Laboratory in Tennessee not only conducts research on nuclear reactors and nuclear fuels but also promotes the development of new materials and innovations in manufacturing technology. Oak Ridge's scientific achievements have been applied not only to the nuclear industry but also to technological advancements in aerospace, automotive manufacturing, and other sectors, driving the development of related industries.

7.1.1.3.2 France

France has secured a significant position in the global nuclear industry through its vigorous development of nuclear power. This advancement has enhanced France's competitiveness in the global energy market while promoting the modernization and technological upgrading of its domestic industries. France's AREVA Group (now renamed Areva) is a world-leading nuclear energy enterprise. Its operations span the entire nuclear fuel cycle, encompassing uranium exploration, fuel processing, nuclear reactor design and construction, and waste management [3]. This integrated industrial chain not only enhances France's competitiveness in nuclear energy but also drives the development of related domestic industries such as mechanical manufacturing, electrical engineering, and chemical industries.

7.1.1.3.3 China

Over the past few decades, China has vigorously developed its nuclear industry, gradually forming a complete nuclear industrial chain. Enterprises, such as China National Nuclear Corporation (CNNC), China General Nuclear Power Group (CGN), and State Power Investment Corporation (SPIC), have not only accumulated extensive experience in nuclear power plant construction and operation but have also made extensive investments in nuclear fuel cycle and nuclear technology applications. China's independently developed third-generation large pressurized water reactors (PWRs), Hualong One and Guohua One, have achieved parity with foreign PWR nuclear power technologies and continue to undergo performance enhancement R&D. The successful application of these reactors signifies China's attainment of self-reliance and industrialization in nuclear power technology. The advancement of the nuclear industry has not only upgraded China's equipment manufacturing sector but also propelled the development of high-tech industries such as nuclear materials and nuclear power equipment, driving rapid growth in related domestic industries.

7.1.2 Nuclear power development

7.1.2.1 Principles and application approaches of nuclear power generation

7.1.2.1.1 Principle features

In a nuclear reactor the nuclei of heavy elements such as uranium or plutonium undergo fission under the bombardment of neutrons, producing two lighter nuclei while releasing a large amount of energy. This energy is used to heat the coolant (water or other liquid) in the reactor, generating high-temperature and high-pressure steam. Steam drives the rotation of the steam turbine, thereby driving the generator to generate electricity. The nuclear fission reaction has a high energy density and can continuously provide a large amount of thermal energy, making it the core of nuclear power generation. Compared with intermittent renewable energy sources such as wind and solar power, nuclear power has higher stability and controllability. Nuclear power plants can operate 24/7 without being limited by weather, seasons, and time and can provide uninterrupted baseload electricity. This stable power supply is particularly important for industrial production, especially for industries that require sustained electricity such as steel smelting, chemical production, and large-scale manufacturing. Nuclear power is an ideal choice.

7.1.2.1.2 Economic advantages

Nuclear power plants have significant advantages in power generation, with extremely high energy density. A small amount of nuclear fuel can generate huge amounts of electricity, greatly reducing fuel transportation and storage capacity; clean and environmentally friendly, the operation process produces almost no pollutants such as carbon dioxide and sulfur dioxide, which is of great significance for improving environmental quality and mitigating climate change; stable and reliable operation, free from excessive interference from natural factors, capable of providing continuous and stable power supply to the power grid; the nuclear fuel resources are relatively abundant, and with the development of technology, the utilization rate continues to improve, which can meet long-term energy needs; and from the perspective of the entire life cycle, although the initial investment is large, the fuel cost is low, the maintenance cost is stable, and the economic benefits are good [4].

7.1.2.1.3 Application requirements

Nuclear power releases energy through nuclear fission reactions and converts this energy into electricity to supply to users. The design and operation of nuclear power plants are based on the fission reaction of heavy elements (such as ^{235}U or ^{239}Pu) inside the nuclear reactor. During the fission process a large amount of heat energy is released to generate high-pressure steam, which drives the steam turbine to generate electricity. In industrial applications the stable output of nuclear power plants can ensure the continuous operation of factories and production facilities, especially for heavy industrial enterprises that have strict requirements for power supply. In addition, artificial intelligence has become a positive growth pole for the new demand for nuclear energy.

7.1.2.2 Nuclear power's role in promoting industrial development

7.1.2.2.1 Reducing energy costs

Nuclear power plants feature low operating costs, especially in fuel expenses. The high energy density of nuclear fuel results in long-term operational costs lower than those of traditional thermal power plants. For energy-intensive industrial enterprises, utilizing nuclear power can significantly lower production costs and enhance product competitiveness in the market.

7.1.2.2.2 Driving industrial upgrading

Nuclear power development requires high-tech equipment and materials, such as nuclear reactors, steam generators, and control systems. The research, development, and manufacturing of these components promote technological advancement in related industries. The implementation of nuclear power projects also stimulates the growth of high-end manufacturing, elevating industrial automation levels and overall industrial capabilities.

7.1.2.2.3 Supporting emissions reduction and environmental protection

As a low-carbon energy source, nuclear power generation produces no carbon dioxide emissions, aiding industrial sectors in achieving energy conservation and emissions reduction targets. In some countries the expansion of nuclear power has effectively reduced reliance on coal-fired power plants, lowering coal consumption and air pollution while providing energy security for the green transformation of industrial sectors.

7.1.2.3 Nuclear power plant development

7.1.2.3.1 Global nuclear power plants

Data from the International Energy Agency indicates that global nuclear power generation is poised to reach new heights. With increased investment, technological advancements, and policy support, the electricity output from nearly 420 nuclear reactors worldwide is expected to further expand. Global nuclear power generation is projected to grow at an annual rate of nearly 3%, reaching a new historical peak by 2026 [5].

7.1.2.3.2 China's nuclear power plants

Nuclear power plants exemplify the application of nuclear technology. As one of the world's fastest growing nuclear energy nations, China operated 51 nuclear reactors by 2024, ranking third globally, while leading the world in the number of reactors under construction. China's domestically developed third-generation nuclear technologies, such as Hualong One and Guohua One, are not only widely deployed domestically but are also expanding into international markets. China has achieved remarkable progress in nuclear energy. By December 2025 the country operated, constructed, and approved a total of 102 nuclear reactors with an installed capacity of 113 million kW. Among these, 28 reactors under construction had a combined capacity of 33.65 million kW, whereas 60 commercial reactors generated 60.96 million kW—the world's largest operational fleet.

7.1.2.4 Practical applications of nuclear power in industrial development across nations

7.1.2.4.1 United States

As one of the earliest adopters of nuclear power globally the United States relies on its nuclear plants to provide substantial baseload electricity to the national grid, supporting numerous industrial sectors, including chemical processing, metallurgy, and manufacturing. Through consistent nuclear power supply the United States has diversified its energy sources, reduced dependence on fossil fuels, and advanced its global leadership in nuclear technology and equipment manufacturing.

7.1.2.4.2 France

France boasts the world's highest proportion of nuclear power, accounting for over 70% of its total electricity generation. This substantial nuclear supply not only meets domestic industrial electricity demands but also underpins France's energy exports. France's nuclear advantage significantly lowers industrial electricity costs, attracting substantial investment from energy-intensive industries to establish facilities within the country. France's nuclear development has also accelerated the growth of its nuclear technology and equipment manufacturing sectors, positioning it as a major player in the global nuclear power market.

7.1.2.4.3 Japan

Japan's nuclear power once provided stable electricity for industry, supporting energy-intensive sectors like steel and chemicals. Following the Fukushima nuclear accident, nuclear power's share declined, though existing reactors continue to supply electricity. In addition, nuclear-derived technologies find applications in medicine (e.g., radioactive isotopes) and industrial inspection, with their industrial chain driving development in materials, machinery, and other sectors.

7.1.2.4.4 South Korea

South Korea successfully met its massive electricity demands during rapid industrialization through nuclear power development. South Korea's nuclear power plants are widely distributed across the country, providing stable electricity supply to key industries such as steel, petrochemicals, and electronics, supporting the sustained growth of these sectors. South Korea's nuclear power technology has also advanced alongside industrial development, enhancing the international competitiveness of domestic nuclear equipment manufacturing and nuclear engineering construction.

7.1.2.4.5 China

China has vigorously developed nuclear power in recent decades to meet the growing electricity demand during its rapid industrialization and urbanization.

China's nuclear power projects—including Qinshan, Daya Bay, Shidaowan, Xiapu, Tianwan, Sanmen, and Zhangzhou nuclear power plants—have significantly expanded the nation's power generation capacity, providing robust electricity support for industries such as steel, automotive, and electronics. China's nuclear power development has also driven the upgrading of its domestic equipment manufacturing sector, creating a vast market for the research, development, and production of high-tech nuclear power equipment and materials.

7.1.2.5 Challenges and prospects of nuclear power in industrial development

7.1.2.5.1 Nuclear power plant safety remains a focal point of public concern

Nuclear power plant safety remains a focal point of public concern. Historical nuclear accidents, such as Chernobyl, Three Mile Island, and Fukushima, have fueled public apprehensions about nuclear power safety. Strengthening safety management at nuclear power plants and enhancing the reliability of nuclear power technology are critical to the further development of nuclear power [6].

7.1.2.5.2 Waste disposal and storage are issues nuclear power development must confront

Nuclear power plants generate radioactive waste during operation, and the disposal and storage of this waste are problems nuclear power development must address. Although current waste treatment technologies continue to advance, safer and more environmentally sound solutions remain necessary [7].

7.1.2.5.3 Reasonable planning is required for energy policies and economic incentives

The construction and operation of nuclear power plants demand substantial upfront investment, involve lengthy project cycles, and are influenced by multifaceted policy and economic factors. Governments worldwide must establish well-structured energy policies and economic incentive measures to ensure the sustainability of nuclear power projects.

7.1.3 Development of nuclear heating

7.1.3.1 Principles and applications of nuclear heating

7.1.3.1.1 Fundamental characteristics

The core principle of nuclear heating involves utilizing thermal energy generated by nuclear reactors. Through a series of heat exchange and transmission processes, this energy supplies heating networks or industrial processes with the required thermal output. Unlike nuclear power plants primarily focused on electricity generation, nuclear heating centers on harnessing thermal energy rather than electricity. This distinction grants nuclear heating unique advantages in energy utilization efficiency and direct thermal supply [8].

7.1.3.1.2 Reactor types and thermal energy transfer mechanisms

Nuclear heating typically employs small- or medium-sized reactors designed for low-temperature, low-pressure operation, ensuring high safety and operational simplicity. Common reactor types include PWRs and heavy water reactors (HWRs). During operation, these reactors generate substantial thermal energy through nuclear fission. This heat is absorbed by a coolant (such as water, gas, or liquid metal) and transferred to a heat exchange system. Through heat exchangers the thermal energy is then conveyed to a secondary coolant loop, ultimately feeding into urban heating networks or industrial thermal systems.

7.1.3.1.3 Application pathways

Nuclear heating finds broad application across various scenarios, including district heating systems, industrial heat supply, seawater desalination, and agricultural greenhouse heating. In district heating, nuclear power can replace traditional coal- or gas-fired boilers to provide stable heating services for large residential areas. Within industrial sectors, nuclear heating is particularly suited for applications demanding high thermal energy volumes and stable heat supply, such as petrochemical processing, steel smelting, and papermaking. In addition, nuclear heating can serve as a reliable energy solution for remote areas inaccessible to conventional power grids.

7.1.3.2 Nuclear heating's contribution to industrial development

7.1.3.2.1 Reducing carbon emissions

Traditional industrial heating relies heavily on coal- or gas-fired boilers, which generate substantial carbon dioxide and pollutants. Nuclear heating produces zero carbon emissions during thermal energy generation, aiding carbon neutrality goals and reducing the industrial sector's carbon footprint.

7.1.3.2.2 Lowering energy costs

Nuclear heating benefits from abundant, stable energy sources with relatively low fuel costs and extended operational lifespans. This significantly reduces industrial energy expenses, with its cost advantages becoming particularly pronounced during periods of volatile energy pricing.

7.1.3.2.3 Enhancing energy utilization efficiency

Through efficient thermal conversion, nuclear heating systems improve energy utilization efficiency and minimize waste. This high-efficiency heating method optimizes industrial thermal systems, boosts production efficiency, and reduces energy consumption per unit of output.

7.1.3.3 Practical applications of nuclear heating in industrial sectors worldwide

7.1.3.3.1 United States

The United States has actively explored and progressively advanced practical applications of nuclear heating in industrial sectors, adopting technological approaches such as low-temperature heating reactors and advanced small modular reactors (SMRs). In its early stages, attempts were made to convert the Trojan Nuclear Power Plant in Oregon into a regional heating system. Although this initiative did not materialize, it yielded valuable experience. The Idaho National Laboratory also conducted experiments to validate the feasibility of coupling reactors with chemical processes. Today, advanced SMRs are the primary focus. NuScale Power's 60 MW SMR has received certification from the US Nuclear Regulatory Commission (NRC) and is slated to supply thermal energy for steam-assisted oil recovery in Utah's oil and gas fields, replacing gas-fired boilers to reduce carbon emissions.

7.1.3.3.2 Russia

Russia possesses abundant nuclear resources and extensive experience in nuclear heating. It has deployed small nuclear reactors in remote Arctic regions to provide stable thermal energy for local industry and residents. This model effectively addresses energy supply challenges in extreme climates while reducing fossil fuel use and environmental pressure.

7.1.3.3.3 Finland

Finland has actively explored multipurpose nuclear applications in recent years. The Loviisa Nuclear Power Plant in Finland utilizes waste heat to supply steam to nearby industrial facilities in addition to generating electricity. This approach not only enhances nuclear power utilization efficiency but also provides local industries with an economical and efficient energy solution.

7.1.3.3.4 China

As the world's largest energy consumer, China has actively pursued nuclear heating initiatives. The Haiyang Nuclear Power Plant in Rongcheng, Shandong Province, marks China's first commercial project utilizing nuclear power for heating, supplying winter warmth to nearby industrial parks and residential areas. This nuclear heating project has significantly reduced coal consumption and carbon dioxide emissions in the region, yielding notable environmental benefits. The stability and cleanliness of nuclear heating make it a crucial tool for China in achieving its "dual carbon" goals. The Tianwan Nuclear Power Base in Lianyungang, Jiangsu Province, will supply heating to an area of 14.02 million m^2 during the 2024–25 heating season through its "Heqi No. 1" heating project—China's first initiative to provide steam to a petrochemical industrial park. The

Xuyi Nuclear Power Project, located in Xuyi County, Huaian City, Jiangsu Province, features a large planned total installed capacity. Utilizing advanced nuclear power technology, it aims to enhance energy efficiency, reduce carbon emissions, and promote regional economic development. On November 21, 2025, China Hainan Nuclear Power achieved its first stable industrial steam supply to adjacent enterprises, marking China's advancement from single-purpose power generation and residential heating to a new phase of comprehensive nuclear energy utilization for industrial steam supply.

7.1.3.4 Challenges and prospects of nuclear heating in industrial development

7.1.3.4.1 Safety concerns

Nuclear safety remains a focal point of public attention. Although heating reactors are smaller in scale and incorporate multiple technical safeguards, enhancing safety further to gain public trust remains a key issue to address during the promotion of nuclear heating.

7.1.3.4.2 Cost issues

Nuclear heating demands significant technological investment across reactor design, construction, and operation. High initial construction costs necessitate policy support and long-term investment from governments and enterprises.

7.1.3.4.3 Policy and regulatory issues

Countries maintain varying policies and regulations governing nuclear energy development. Promoting nuclear heating requires compliance with international safety standards and national policies. Robust policy frameworks and regulatory safeguards will be pivotal to the successful implementation of nuclear heating projects. Against the backdrop of the global energy transition, pursuing efficient, clean, and stable energy supply methods has become a key driver for sustainable industrial development.

7.1.3.4.4 New technological solutions for future energy transition

As an emerging and highly promising energy supply form, nuclear heating is gradually gaining prominence in industrial sectors, exerting multidimensional positive impacts on industrial development. Looking ahead, with continuous advancements in nuclear technology and growing demand for clean energy, nuclear heating holds broad application prospects in industrial sectors. The development of new nuclear technologies, such as SMRs, will further enhance the safety and economic viability of nuclear heating, driving its adoption across more countries and regions [9]. Nuclear heating not only contributes to achieving low-carbon industrial development goals but also offers new solutions for the global energy transition.

7.1.4 Nuclear power development

7.1.4.1 Principles and applications of nuclear power

7.1.4.1.1 Key characteristics

The core principle of nuclear power involves converting thermal energy released by nuclear fission reactions into motive power: using fissile isotopes such as ^{235}U as fuel, neutrons bombard the nuclei of nuclear fuel atoms within a reactor, triggering fission and releasing substantial thermal energy; this heat warms the coolant (e.g., light water, heavy water, liquid metal), generating high-temperature, high-pressure steam or superheated fluid. This steam then drives turbines to turn generators for electricity production (nuclear power generation) or directly powers propulsion systems (e.g., ship propellers), converting thermal energy into mechanical/electrical energy. Nuclear fission boasts extremely high energy density, enabling sustained power generation with minimal fuel.

7.1.4.1.2 Application pathways

7.1.4.1.2.1 Energy supply pathway Used for nuclear power generation to meet large-scale, stable electricity demands, particularly suited for base-load power supply. It compensates for the intermittency of renewable sources such as wind and solar power, supporting energy structure transformation under the "dual carbon" goals.

7.1.4.1.2.2 Specialized transportation pathway Applied in nuclear-powered vessels such as submarines, aircraft carriers, and icebreakers. Leveraging its long endurance, infrequent refueling requirements, and high-power characteristics, it supports extended-duration, high-load missions such as deep-sea navigation and polar expeditions.

7.1.4.1.2.3 Power supply for specific scenarios Provides independent, sustained energy for remote regions, islands, spacecraft, and submersibles—including future nuclear-powered satellites—addressing challenges in conventional energy transportation or supply shortages.

7.1.4.2 Nuclear power's role in industrial advancement

7.1.4.2.1 Providing stable, efficient energy

Nuclear power offers exceptional energy density and endurance, delivering continuous electricity to energy-intensive industries like steel and chemicals. It ensures production continuity unaffected by fossil fuel supply fluctuations, particularly suited for large industrial hubs with concentrated energy demands.

7.1.4.2.2 Supporting specialized industrial scenarios

In remote areas, islands, and other locations lacking grid coverage—such as offshore oil and gas extraction or polar resource exploration—nuclear power can function as an independent energy system. It meets equipment operational and personnel living requirements, expanding the scope of industrial activities.

7.1.4.2.3 Driving industrial technology upgrades

Nuclear power applications stimulate upstream and downstream industries, including nuclear fuel cycles, nuclear equipment manufacturing, and nuclear safety monitoring. This drives technological innovation in high-end equipment, materials science, and automated control systems, forming a complete industrial chain and enhancing overall industrial technological competitiveness.

7.1.4.3 National development

7.1.4.3.1 United States

7.1.4.3.1.1 Early 1950s to 1990s The United States pioneered research into nuclear electric propulsion for space in the 1950s. Among these efforts the 100-W-class isotope thermoelectric generator utilized high-temperature SiGe materials, achieving an electrical output of 170 W with a hot-end temperature exceeding 1000°C. It was successfully deployed on the Voyager 1 spacecraft, operating for over 43 years. Although such space isotope power sources provide relatively modest electrical output, nuclear propulsion requires harnessing nuclear fission/fusion reactions within reactors. Through thermoelectric conversion, significantly greater electrical power can be generated. In 1955 the US Atomic Energy Commission launched the Space Nuclear Auxiliary Power (SNAP) program. In April 1965, it successfully launched the first operational space nuclear reactor power source in human history—SNAP-10A. Employing thermocouple power generation based on temperature differences, it supplied 500 W of electrical power to spacecraft before being permanently deactivated after 43 days of operation. The 1979 Three Mile Island nuclear accident impacted the US nuclear industry, leading to a reduction in new nuclear power plant construction. From the 1980s to the 1990s the United States continued developing advanced nuclear reactors. General Electric developed the advanced boiling water reactor (ABWR), whereas Westinghouse developed the AP600 passive safety PWR. Concurrently, nuclear-powered aircraft carrier technology advanced, with the Nimitz-class nuclear-powered carriers becoming the backbone of the US Navy.

7.1.4.3.1.2 Since the 21st century In July 2015 the United States released the detailed NASA Technology Roadmap. Regarding nuclear propulsion technology, it categorized space nuclear reactor power sources into three power tiers: 1–10 kW: applications: power sources for scientific mission buses and manned planetary exploration spacecraft; 10–100 kW: flexible path missions like crewed asteroid exploration; 1–5 MW: crewed Mars exploration missions requiring low mass density (<5 kg/kW). In NASA's 2020 Technology Classification, both the propulsion systems and space power and energy storage categories emphasized space nuclear propulsion and space nuclear reactor power sources. The primary nuclear technology deployed by the United States in space is radioisotope power sources. For instance the Perseverance rover, launched on July 30, 2020, is powered by a multimission radioisotope thermoelectric generator (RTG). The US space nuclear power program, Kilopower, targets power generation capacities of 1–10 kW. In March 2018 the 1-kW demonstration unit of the Kilowatt Reactor Using Stirling Technology (KRUST) project achieved success. By 2025 the US Department of Energy's Office of Clean Energy Demonstrations will select and fund two advanced small light-water reactor demonstration projects to mitigate deployment risks for advanced reactors. Regarding nuclear-powered vessels, by 2025, the United States operates 70 nuclear-powered ships, including 50 Los Angeles-class and Virginia-class submarines. The US Nimitz-class aircraft carrier is equipped with two A4W reactors, delivering a combined power of approximately 200 MW to propel its 300,000-t hull at speeds exceeding 30 knots. In addition, in the field of space nuclear propulsion, the United States is actively pursuing research and development, such as NASA's DRACO project, which aims to achieve power outputs of 10–20 MW and plans to conduct tests by 2027.

7.1.4.3.2 Russia

7.1.4.3.2.1 Early 1950s to 1990s Beginning in the 1950s the former Soviet Union concurrently conducted extensive and in-depth research into space nuclear propulsion. Power systems like the BUK thermoelectric generator, utilizing the thermoelectric conversion principle, achieved dozens of successful in-orbit deployments aboard the "Cosmos" series reconnaissance satellites. Building on this foundation, the USSR also successfully launched the TOPAZ-I thermionic conversion nuclear power source, achieving in-orbit operation. Furthermore, it completed the development and ground testing of a full-scale prototype for the TOPAZ-II nuclear power source, accumulating substantial technical data for the advancement of space nuclear power. Following the dissolution of the Soviet Union, related research entered a low phase because of insufficient funding. In the 1990s the Russian government passed a resolution to develop space nuclear power sources, deciding to further advance high-power, long-life thermionic space reactor power systems.

7.1.4.3.2.2 Since the beginning of the 21st century Since the 21st century the national strategy focused on developing deep space exploration has prompted Russia to resume research on space nuclear propulsion. On April 24, 2008 the Russian government approved the "Principles of the Policy of the Russian Federation in the Field of Space Activities until 2020 and Beyond," indicating the government's significant demand for comprehensive space research, exploration, and utilization. Then-President Dmitry Medvedev approved a space nuclear propulsion system program worth 17 billion rubles. Following technical evaluations, Russian authorities concluded that nuclear thermal propulsion development was prohibitively expensive with limited practical applications. Conversely, advancements in electric propulsion technology enhanced the feasibility of nuclear electric propulsion concepts. Consequently, Russia decided to concentrate efforts on developing high-power thermoelectric conversion technology, initiating extensive international cooperation with European Union nations. In September 2020 the Arsenal Design Bureau (KBArsenal) commenced assembly of the nuclear-powered spacecraft, aiming to launch the first nuclear-powered space tug into orbit by 2030 for flight testing. According to the mission plan: In the second phase the tug will continue toward Venus, conducting xenon refueling tests en route while releasing a research satellite from its payload module. The tug itself, along with remaining scientific equipment, will perform gravity assists to enter the third phase: a mission to Jupiter's moons for scientific study. Since assembly commenced in September 2020, Russia's nuclear-powered spacecraft project has advanced steadily. Despite challenges, its prospects remain promising. Should the first nuclear-powered space tug successfully complete flight tests before 2030, Russia will undoubtedly lead the global nuclear-powered spaceflight field, blazing new trails for human space exploration and propelling our understanding and development of the cosmos to unprecedented heights. In September 2025, Russia announced the successful development of two high-power propulsion systems. One is the KM-50M Hall thruster employing shielding technology; the other is the ID-750 ion thruster utilizing plasma technology. In September 2025 the second RITM-400 reactor for the world's most powerful nuclear-powered icebreaker, the Rossiya, completed trial installation. This reactor's thermal power was increased to 315 MW, enabling the Rossiya to break through ice up to 4 m thick. In October 2025 the nuclear-powered cruise missile "Burevestnik" was test-fired, and the nuclear-powered underwater unmanned submersible "Poseidon" was successfully developed.

7.1.4.3.3 France

7.1.4.3.3.1 Early 1950s to 1990s France launched the "Phoenix Program" in 1956, focusing on fast neutron reactor development. In 1967, it completed the experimental fast reactor "Rhapsody," becoming one of the few nations globally to master breeder reactor technology. Simultaneously, to address power shortages, France signed an agreement with Westinghouse in 1957 to introduce PWR technology. Construction began on its first commercial nuclear power plant, the Schutz plant, in 1962. The commissioning of the Chinon A1 plant (90 MW) in 1967 marked the beginning of France's nuclear self-sufficiency. In 1969, Électricité de France (EDF) and Framatome jointly promoted the localization of PWR technology, laying the foundation for the "Unified Reactor Strategy" of the 1970s. In the 1980s, nuclear power plants were massively expanded using PWR technology, rapidly increasing nuclear's share to become the primary energy source; the 1990s further consolidated nuclear's dominant position, refined the nuclear fuel cycle system, and simultaneously began prioritizing nuclear safety and decommissioning technology R&D, transitioning the nuclear industry from rapid expansion to stable operation. After initial setbacks in nuclear submarine development during the 1950s, France resolved to pursue independent research in the 1960s. Starting in the 1970s, the red-letter class strategic nuclear submarines and the Rubis class attack nuclear submarines were successively launched and continuously upgraded. Regarding aircraft carriers, France enhanced its capabilities by leasing foreign carriers in the early 1950s and began developing its own Clementose-class conventionally powered carriers by the late 1950s. In the 1980s, France commenced development of the nuclear-powered aircraft carrier Charles de Gaulle. By the 1990s the Clemenceau-class carriers were gradually decommissioned, and the Charles de Gaulle was launched in 1994.

7.1.4.3.3.2 Since the 21st century France's nuclear power development has undergone policy adjustments and technological advancements. Initially influenced by factors such as the Fukushima nuclear accident, nuclear power development slowed, and France planned to reduce nuclear energy's share in its energy mix. In 2021, to achieve energy independence and combat climate change, France restarted its nuclear energy development program. In 2024, Unit 2 of the Creyas nuclear power plant in France became the world's first nuclear power unit to process and recover uranium fuel from a fully loaded core, marking a milestone in the development of a closed fuel cycle. Concurrently, France is actively developing small reactors, such as the Nuward small reactor, to meet diverse needs, including power supply for remote areas. Furthermore, in 2023, the French government set deadlines for decision-making and preparatory work on new European Pressurized Reactor-2 (EPR-2) units and spent fuel reprocessing facilities, vigorously advancing nuclear power development to consolidate its global leadership in nuclear technology. France plans to construct six next-generation EPR reactors by 2050, aiming to increase nuclear power generation capacity by 25 GW, while conducting feasibility studies for an additional eight reactors. Regarding nuclear-powered submarines, France currently operates four Triomphant-class ballistic missile submarines. Planning for a third-generation ballistic missile submarine began in 2012, with the project formally approved in 2017. The steel-cutting ceremony for the first vessel took place in March 2024, with delivery of the first submarine expected after 2035. Four vessels will be built, focusing on enhancing stealth, maneuverability, and automation capabilities. Regarding aircraft carriers the nuclear-powered Charles de Gaulle entered service in 2001. France proposed a "new generation aircraft carrier" program in 2018, with construction officially confirmed in 2020. This nuclear-powered, medium-sized carrier will have a full load displacement of 75,000–80,000 t and is expected to enter service in 2037. In 2025, France's Nuclear

Safety Authority approved EDF's use of four MOX fuel assemblies in Unit 4 at the Paluel Nuclear Power Plant. This marks France's first deployment of MOX fuel assemblies in a 1300-MWe reactor unit.

7.1.4.3.4 China

7.1.4.3.4.1 Early 1950s to 1990s The application of nuclear power in China is mainly concentrated in nuclear power generation, nuclear-powered ships, and other industrial and scientific research fields. PWRs use water as a coolant and moderator and have the advantages of compact structure, high power density, and good safety. HWRs use heavy water as a moderator, and their characteristic is that they can use natural uranium as fuel and can continuously refuel during operation. Canada's CANDU reactor type is a representative of HWRs, and China has also introduced and applied related technologies. In 1985 the first phase of China's self-designed and constructed Qinshan Nuclear Power Station began construction, marking a breakthrough in mainland nuclear power from scratch. At the same time the Daya Bay Nuclear Power Station, which introduced technology, began construction, laying the foundation for the development of nuclear power. In 1991, Qinshan Phase I and Daya Bay Nuclear Power Station were successively put into operation, achieving commercial operation of nuclear power. In terms of nuclear-powered submarines, China initiated the development of nuclear submarines in 1958. The first attack-type nuclear submarine, the Long March 1, was launched in 1970 and officially commissioned in 1974, marking China as the 5th country to possess nuclear submarines. Subsequently a strategic nuclear submarine was developed, and in 1988, it successfully completed underwater launch of submarine-to-ground missiles, possessing the capability for a second nuclear strike. In the field of aircraft carriers, this stage mainly focuses on technological exploration and theoretical research, without actual construction of aircraft carriers, but lays the foundation for the development of aircraft carriers after the 21st century.

7.1.4.3.4.2 Since the beginning of the 21st century China is actively developing and demonstrating more advanced reactor types. The "Hualong One" is a major landmark achievement in China's innovative development of nuclear power, with numerous independent intellectual property rights and more than 700 patents and 120 software copyrights. Its research and application have driven technological innovation and upgrading in related industries, enhanced China's technological level in high-end equipment manufacturing, materials science, and other fields, and promoted the overall development and progress of the nuclear power industry chain. Guohe No. 1 is a third-generation nuclear power independent landmark achievement developed based on the national major science and technology special project "Large Advanced Pressurized Water Reactor and High-Temperature Gas-Cooled Reactor Nuclear Power Plant". It is a completely independently designed Chinese nuclear power technology brand, representing the advanced level of nuclear power technology today. There is also a high-temperature gas-cooled reactor, which uses ceramic-coated granular fuel and helium as a coolant. It has inherent advantages such as good safety and high outlet temperature and can be used for high-temperature process heating. The Huaneng Shidao Bay High-Temperature Gas-Cooled Reactor Nuclear Power Plant is the world's first ball bed modular high-temperature gas-cooled reactor nuclear power plant, marking China's world-leading level in the field of high-temperature gas-cooled reactor technology. In addition, research on fast neutron reactors in China is progressing smoothly. The China Experimental Fast Reactor has been built and has achieved stable operation at full power. The Xiapu Demonstration Fast Reactor has also been completed and put into operation; the design scheme of the sodium-cooled fast neutron high-throughput research reactor has passed the review and is moving toward commercial and large-scale application. Fast reactors can more effectively utilize uranium resources and achieve the proliferation of nuclear fuel, which is of great significance for ensuring the sustainable development of China's nuclear energy. China's nuclear submarines are accelerating iteration, and the new generation of strategic nuclear submarines has entered the development stage. China's new generation 09 series nuclear submarines are equipped with advanced antiship missiles, antisubmarine missiles, and vertical launch systems, becoming a key force for the Chinese Navy to move toward deep blue. China is building nuclear-powered icebreakers. Nuclear-powered icebreakers have strong power output and sustained navigation capabilities, which can open up channels in thick ice sea areas in extremely cold regions, providing strong support for polar scientific research, resource development, and other activities. Compared with traditional icebreakers, nuclear-powered icebreakers do not require frequent refueling, reducing reliance on logistical supplies and improving operational efficiency and safety in polar regions. China is also advancing the construction of the world's first nuclear-powered container ship. The construction of China's nuclear aircraft carrier is an important component of the development of the country's national defense forces. Relevant work is steadily advancing in accordance with the needs of national defense construction, aiming to enhance the navy's ability to engage in long-range operations and the country's strategic defense level. Its development always follows the laws of national defense construction and the overall development plan of the country. China has launched the construction of its first nuclear-powered aircraft carrier in 2025. In 2025, China has also approved the construction of 10 new nuclear power plants, of which 8 will use "Hualong One" technology and 2 will use CAP1000 technology. In addition, China has also made progress in the field of supercritical carbon dioxide power generation technology. On November 9, 2025 the world's first commercial supercritical carbon dioxide power generation unit was successfully connected to the grid.

7.1.4.4 Challenges and prospects of nuclear power in industrial development

7.1.4.4.1 Difficulties in waste disposal

The disposal of spent fuel and high-level radioactive waste lacks a clear global strategy, and although geological disposal is technically feasible, the public has doubts about its safety. The increasing amount of waste has become a key obstacle to the development of nuclear power in terms of safe and long-term disposal.

7.1.4.4.2 Talent shortage dilemma

The knowledge base of the nuclear industry is facing the problem of personnel turnover caused by retirement, insufficient new graduates entering the nuclear-related field, and insufficient talent reserves, which pose a potential threat to the sustainable development of the nuclear power industry [10].

7.1.4.4.3 Meet the needs of special scenarios

In deep waters, remote areas, islands, or areas with weak power grids, nuclear power can serve as an independent and reliable energy system to provide energy for industrial activities. For example, space exploration, deep-sea development, offshore oil and gas extraction, and polar resource exploration expand the geographical scope of industrial development.

7.2 Nuclear technology and agricultural development

7.2.1 Crop gene optimization

7.2.1.1 The purpose of crop genetic optimization

This technology can generate new genetic variations, thereby aiding in the development of new varieties with specific traits. Radiation breeding is a method of mutagenic breeding that uses ionizing radiation to treat organisms, inducing mutations from which superior variants are selected. Through a series of breeding procedures, new varieties are cultivated, representing a significant application of nuclear technology in agriculture. Radiation breeding breaks existing genetic chains and overcomes hybrid incompatibility in distant crosses. It facilitates genetic recombination, increases mutation frequency, broadens mutation scope, ensures mutation stability, and shortens breeding cycles. As the global population grows, demand for food continues to rise. The application of nuclear technology in crop gene optimization provides an effective solution to meet this demand. Simultaneously, by reducing production costs, it enhances farmers' economic returns and promotes sustainable agricultural development. The agricultural research community has achieved significant results in nuclear technology applications, including the creation of mutant germplasm resources and the development of new crop varieties. These advancements play a unique role in safeguarding national food security and advancing green agricultural development.

7.2.1.2 Principles of mutation breeding technology

7.2.1.2.1 Definition

Mutation breeding is a technique that utilizes physical or chemical factors (such as radiation or chemicals) to induce genetic mutations in plants. In nuclear technology, gamma rays, X-rays, or neutron beams are the most commonly used mutagens.

7.2.1.2.2 Methods

7.2.1.2.2.1 Mutation induction Exposing crop seeds or tissues to specific radiation doses to induce genetic mutations.

7.2.1.2.2.2 Cultivation of mutant populations Planting irradiated seeds to cultivate mutant populations.

7.2.1.2.2.3 Screening and selection Within the mutant population, screen for plants exhibiting desired nutritional traits and subject them to further cultivation and selection.

7.2.1.2.2.4 Performance testing Conduct nutritional analysis on selected plants to ensure they meet improvement objectives.

7.2.1.2.2.5 Variety stabilization and evaluation Through multiple generations of screening and cultivation, ensure the genetic stability of new varieties and assess their field performance.

7.2.1.3 Specific applications

7.2.1.3.1 Boosting US agricultural yields and reducing costs

Nuclear technology applications in agriculture have opened new avenues for increasing crop yields and reducing production costs. In the United States the development of "Louis" soft wheat represents a successful case study of nuclear technology application. Through radiation mutagenesis, scientists created wheat varieties characterized by early maturity and high yield potential. This wheat not only adapts to diverse climatic conditions but also exhibits a shortened growth cycle, enabling farmers to achieve greater output in less time. Furthermore, its reduced requirements for fertilizers and water resources further cut cultivation costs, thereby lowering human food expenditures.

7.2.1.3.2 Boosting Italian agricultural yields and reducing costs

Italian durum wheat has also benefited from nuclear technology. This wheat variety not only exhibits enhanced disease resistance but also improved adaptability to adverse environments such as drought and saline-alkali conditions. Nuclear technology–enhanced wheat can thrive in poor soil conditions, reducing reliance on chemical fertilizers and pesticides. This not only lowers production costs but also contributes to environmental protection.

7.2.1.3.3 Boosting Soviet agricultural yields and quality

The Soviet Union's "Novosibirsk 67" wheat variety is renowned for its early maturity and high productivity. The application of nuclear technology further optimized these traits, significantly improving both wheat yield and quality.

7.2.1.3.4 Boosting China's agricultural production and cost reduction

7.2.1.3.4.1 In wheat breeding The Institute of Atomic Energy Utilization at Shandong Academy of Agricultural Sciences and the Institute of Crop Science at Chinese Academy of Agricultural Sciences combined space mutagenesis with conventional breeding to develop the new wheat variety "Lu Yuan 502," characterized by broad adaptability, high yield, and stable production. This breakthrough addressed the production challenge of lodging in heavy-spiked varieties, with cumulative adoption reaching 77 million mu (approximately 5.1 million ha). By 2025, it became China's second-most widely promoted wheat variety. The Jiangsu Lixiahe Agricultural Science Research Institute integrated radiation mutagenesis with conventional breeding to develop the "Yangfu Mai 3" series and other new varieties, achieving cumulative adoption across 30 million mu.

7.2.1.3.4.2 Rice breeding The Sichuan Institute of Atomic Energy combined radiation mutagenesis with hybrid vigor utilization technology in indica–japonica crosses, creating new rice germplasm with strong recovery ability, high compatibility, and excellent disease resistance. It developed high-yielding, disease-resistant interspecific hybrid rice varieties like "II You D069," which have been promoted across over 1.4 million acres. The Institute of Nuclear Agriculture and Space Breeding at Hunan Academy of Agricultural Sciences developed the new late-maturing hybrid rice variety "C Liangyou 266," which has been promoted across over 4.7 million mu. The development and application of these improved varieties not only enhance agricultural productivity but also contribute to addressing global food security challenges.

7.2.2 Preservation of agricultural products

7.2.2.1 Purpose of nuclear technology preservation

The application of nuclear technology in agriculture and animal husbandry represents a significant domain within civilian nuclear technology utilization. It will become a vital component of modern agricultural science and technology development, as well as the establishment of green agricultural and animal husbandry brands. The provision of electron beam irradiation preservation services for agricultural products is not only a technological innovation but also a crucial measure to advance agricultural industrial modernization and increase farmers' income. With technological advancements and broader adoption an increasing variety of agricultural products are expected to benefit from this modern preservation technique, achieving more efficient storage and extended shelf life.

7.2.2.2 Principles of low-temperature electron beam irradiation preservation

This process utilizes the physical effects generated by the interaction of electrons—produced by an electron accelerator—with materials to treat agricultural products, thereby effectively extending their shelf life. When radiation penetrates food, it ionizes or excites biomolecules such as DNA and proteins within microbial cells, disrupting their genetic material and metabolic functions to inhibit microbial reproduction. Simultaneously, radiation interferes with enzyme activity in food, slowing respiration and ripening processes in fruits and vegetables, thereby extending shelf life with minimal alteration to nutritional content or sensory qualities. This harmless treatment method offers advantages such as no radiation residue, rapid processing speed, broad applicability, and excellent results, maximizing the preservation of agricultural products' color, aroma, and taste.

7.2.2.3 Specific applications

7.2.2.3.1 Establishment of Preservation Demonstration Center

China has made significant progress in agricultural product storage and preservation, particularly in utilizing nuclear technology for agricultural preservation. The establishment of China's first electron beam irradiation preservation demonstration center at the agricultural product origin marks a significant stride in applying technological innovation to agriculture. Located in Baise City, Guangxi Zhuang Autonomous Region, the center's operation has delivered tangible benefits to farmers within a 200-km radius. This demonstration center employs low-temperature electron beam irradiation preservation technology.

7.2.2.3.2 Red-fleshed kiwifruit preservation

Taking Baise's specialty—organic red-fleshed kiwifruit—as an example, irradiation treatment extends its shelf life from 40 to 100 days. This significantly lengthens the sales period for agricultural products, enabling staggered sales and helping farmers increase their income. In addition, the application of electron beam irradiation technology has reduced the incidence of spoilage in stored mangoes by 30%–50%. This not only minimizes fruit loss but also supports the cultivation and promotion of specialty agricultural products, driving the stable development of distinctive agriculture.

7.2.2.3.3 Potato preservation

As a major potato-growing region, Inner Mongolia traditionally relied on rudimentary storage cellars. With spring warming, potatoes often sprouted or rotted, causing economic losses for farmers. The introduction of electron beam irradiation preservation technology offers a new solution for preserving potatoes and other agricultural products, reducing farmers' economic losses to some extent and enhancing the economic efficiency of the potato industry.

7.2.2.3.4 Enhancing aroma in irradiated liquor

Nuclear radiation processing of liquor involves exposing newly distilled baijiu to electron beams or gamma rays. This generates abundant active free radicals in water molecules, accelerating oxidation and esterification reactions. Within hours to days, newly distilled liquor achieves the quality equivalent to 5–10 years of traditional natural aging. This process not only intensifies the aroma and enriches the palate but also degrades harmful substances like fusel oils. The irradiation occurs in a sealed environment without direct contact with the radiation source, typically at very low doses. This approach complies with food safety standards, eliminating concerns about radioactive contamination residues or induced radioactivity.

7.2.3 Crop fungicide and insecticide

7.2.3.1 Objectives of nuclear technology irradiation for pest control

The application of nuclear technology in crop sterilization and pest control provides modern agricultural production with an efficient and environment-friendly solution. Electron beam irradiation technology enables deep sterilization and pest control of agricultural products without damaging packaging or altering their original quality. This reduces losses from pest infestations and spoilage, eases the workload of agricultural workers, enhances sterilization and pest control efficiency, lowers production costs, and ultimately improves food security for humanity.

7.2.3.2 Principles of nuclear technology irradiation for sterilization and pest control

Electron beam irradiation preservation technology operates by leveraging the ionizing effects of radiation to directly target pests or microorganisms. Radiation damages the reproductive cells of pests, disrupts their developmental cycles, and renders them incapable of reproduction. For microorganisms, radiation causes double-strand breaks in DNA and protein denaturation, eliminating their ability to divide and survive. In addition, low-dose radiation inhibits insect molting and emergence, whereas high doses cause direct lethal effects. This technology offers advantages such as no chemical residues, high processing efficiency, and applicability postpackaging, making it widely used for pest control in grains, dried goods, and similar products.

7.2.3.3 Specific applications

7.2.3.3.1 Fruit fly sterilization

Exposing fruit fly pupae or adults to low-dose gamma rays or X-rays damages their germ cell chromosomes, rendering them incapable of reproduction. Large numbers of sterile males are then released into the wild to mate with wild females. As no viable offspring can be produced, fruit fly populations gradually diminish. This achieves environment-friendly control of agricultural pests like the citrus fruit fly and Mediterranean fruit fly, avoiding chemical pesticide pollution while preserving ecosystems. Fundamental theories and methodologies for radiation-induced sterility in pests have been established. Fujian Agriculture and Forestry University's long-term research on irradiated fruit fly sterility has completed foundational applied studies.

7.2.3.3.2 Sterile mosquito release

Building on years of research, Xi Zhiyong's team at Sun Yat-sen University published a paper in Nature in 2019 reporting that combining cytoplasmic incompatibility technology (IIT) based on the symbiotic bacterium *Wolbachia* with radiation-based insect sterilization technology (SIT) can successfully suppress and eliminate wild *Aedes albopictus* populations on islands. This confirmed the feasibility of the technology and its potential for establishing human-protected areas free from mosquito-borne disease transmission or mosquito nuisance. In 2020, Xi Zhiyong's team achieved a significant breakthrough in the emerging field of using *Wolbachia* to control agricultural pests. Their related research, published in Current Biology, marked the official launch of this new frontier in *Wolbachia*-based agricultural pest control.

7.3 Nuclear technology and transportation development

7.3.1 Nuclear technology applications and development in space

7.3.1.1 Nuclear propulsion systems

7.3.1.1.1 Characteristics

As humanity increasingly prioritizes deep space exploration, the limitations of traditional chemical propulsion systems in endurance and efficiency become increasingly apparent. Two nuclear propulsion system approaches—nuclear thermal propulsion (NTP) and nuclear electric propulsion (NEP)—offer the potential to overcome these bottlenecks. These systems boast advantages such as high specific impulse, high power density, and extended operational lifespans, making them pivotal for future deep space missions.

7.3.1.1.2 Types

7.3.1.1.2.1 Nuclear thermal propulsion (NTP) NTP systems utilize nuclear reactors to heat propellants like hydrogen, achieving specific impulse superior to chemical propulsion. High-temperature gas is expelled through nozzles to propel spacecraft. Systems like the Nuclear Engine for Rocket Vehicle Applications (NERVA) project have demonstrated potential for high energy density and extended endurance.

7.3.1.1.2.2 Nuclear electric propulsion (NEP) NEP utilizes nuclear reactors to generate electricity, powering ion thrusters or electromagnetic propulsion systems. Although offering lower thrust, these systems deliver exceptionally high specific impulse and sustained operational capability, making them ideal for deep-space missions requiring continuous propulsion, particularly in environments distant from the Sun.

7.3.1.1.3 Applications and directions

NASA's NEP program is developing fission-based ion thrusters to support deep-space exploration. In September 2025, Russia announced the successful development of two high-power thrusters: the KM-50M Hall thruster employing shielding technology and the ID-750 ion thruster utilizing plasma technology. Recent advancements in nuclear propulsion technology include optimized reactor designs and improved materials science. Researchers are focused on developing more efficient fuels and higher temperature-resistant materials to enhance the performance and safety of nuclear propulsion systems while also pursuing more compact and lightweight designs suitable for long-duration space missions.

7.3.1.2 Energy supply

7.3.1.2.1 Characteristics

In space exploration, nuclear technology is crucial for energy supply. Its stable, long-term power generation makes it particularly suitable for long-distance exploration and deep-space missions. Nuclear technology primarily provides two energy solutions: fission power system (FPS) and radioisotope thermoelectric generators (RTG).

7.3.1.2.2 Types

7.3.1.2.2.1 Fission power system (FPS) Nuclear power systems generate electricity using nuclear reactors, making them suitable for long-duration deep-space missions. Unlike solar panels, FPS systems do not rely on solar radiation and can operate reliably in environments far from the Sun.

7.3.1.2.2 Radioisotope thermoelectric generator (RTG) RTGs harness the decay of radioactive isotopes (such as plutonium-238) to generate heat, which is then converted into electricity via a thermoelectric generator.

7.3.1.2.3 Applications and directions

RTGs are extensively deployed in deep-space probes such as the Voyager spacecraft and Curiosity rover because of their high reliability and longevity, providing stable power under extreme conditions. Enhanced nuclear power designs and novel materials have improved system efficiency and safety, whereas advancements in isotopes and thermoelectric materials have boosted RTG performance. These technological strides not only enhance energy supply efficiency but also support sustained capabilities during extended missions. Artificial intelligence, leveraging its robust data processing and analytical capabilities, intelligent algorithm optimization, and autonomous learning mechanisms, is deeply integrated throughout the entire energy chain—from production and transmission to storage and consumption. This enables efficient scheduling, precise forecasting, intelligent operation and maintenance, and optimized management of energy systems, powerfully driving the transformation and upgrading of the energy sector toward green, low-carbon, safe, reliable, smart, and efficient development.

7.3.1.3 Energy foundations for lunar development

7.3.1.3.1 Characteristics

Helium-3 is a clean, highly efficient nuclear fusion fuel. Successful extraction and utilization of lunar helium-3 would provide Earth and future lunar bases with a long-term, stable, and highly efficient energy source.

7.3.1.3.2 Resources

The Moon possesses abundant nuclear energy resources. Lunar regolith contains significant quantities of helium-3, with estimated reserves ranging from 1 to 5 million t. Compared to the deuterium–tritium fusion commonly used on Earth, helium-3-deuterium fusion produces fewer neutrons, resulting in lower radioactive contamination while releasing enormous energy.

7.3.1.3.3 Applications and directions

Utilizing nuclear fuels like uranium on the Moon, nuclear reactors can be developed to provide stable power for lunar bases. For example, China's newly designed lunar nuclear reactor, with a power output of 40 kW and a 10-year lifespan as its basic specifications, can meet the electricity demands of life support systems, scientific research equipment, and communication devices at lunar bases. The United States leads in lunar nuclear energy development. Its Lunar Fission Power Reactor Project is a key component of the Artemis program, with Phase 1 completed in 2024. Phase 2 bidding is scheduled to commence in 2025, aiming to deliver ground and flight prototypes by late 2028. These prototypes will feature a design power of 40 kW and be capable of continuous operation on the lunar surface for at least 10 years. In September 2025, Russia successfully developed two new thrusters—the KM-50M Hall thruster and the ID-750 ion thruster. These are reportedly capable of producing helium-3 on the moon and could be used to develop commercial applications, including space tourism.

7.3.2 Application and development of nuclear technology in the ocean

7.3.2.1 Type

The application of nuclear technology in civilian marine transportation is gaining increasing attention, primarily focused on nuclear-powered merchant ships and nuclear-powered icebreakers.

7.3.2.2 Nuclear-powered merchant ships

The development of nuclear-powered merchant ships aims to provide long-term, efficient transportation solutions through nuclear propulsion systems. The emergence of SMRs has made the construction and operation of nuclear-powered vessels more economical and feasible. These reactors not only offer high safety standards but also allow flexible power output adjustment based on demand, positioning them as a key future direction for maritime transportation.

7.3.2.3 Nuclear-powered icebreakers

Nuclear technology also plays a critical role in nuclear-powered icebreakers, which are essential for navigation in polar regions. These vessels can open routes through thick ice layers that would otherwise be impassable. Technical advancements in nuclear-powered icebreakers primarily focus on enhancing reactor power and durability.

7.3.2.4 Examples

Russia's nuclear-powered icebreaker "Arktik" is equipped with two RITM-200 nuclear reactors, delivering a combined output of 60 MW. It can break through ice up to 3 m thick and operate for 7 years on a single fuel load. November 2025, Wuwei, China: As the world's only molten salt reactor incorporating thorium fuel, it provides robust technical support for the design of the "KUN-24AP"—the world's first 24,000 TEU nuclear-powered container ship launched by China's Jiangnan Shipyard in 2023.

7.3.3 Land-based nuclear technology applications and development

7.3.3.1 Nuclear-powered trains

7.3.3.1.1 Concept of nuclear-powered train

The concept of nuclear-powered trains began in the mid-20th century, when scientists hoped to use the high energy density of nuclear energy to propel trains forward. However, because of technological, economic, and security challenges, early plans were not realized. After 2022 the advancement of SMR technology has made nuclear-powered trains a new focus of attention. SMR has a small volume and high safety, making it suitable for land transportation applications.

7.3.3.1.2 Russian nuclear–powered train

Russia has shown a strong interest in nuclear-powered trains. In 2020, Rosatom announced that it was developing SMR-based nuclear-powered trains for long-term freight transportation in polar regions. This type of train can operate in extreme environments for months without the need for external fuel supply.

7.3.3.1.3 Characteristics and safety

Despite the potential for long-term operation of nuclear-powered trains, they still face technical challenges such as the effectiveness and safety of nuclear reactors, the reliability of cooling systems, and public acceptance of nuclear energy. In the future, it is necessary to further develop more efficient and safe nuclear power systems and address related social and policy issues.

7.3.3.2 Nuclear-powered freight vehicles and trucks

7.3.3.2.1 Concept of nuclear-powered freight vehicles

The development of nuclear-powered freight trucks and lorries aims to achieve long-term, efficient, and environment-friendly cargo transportation. Similar to nuclear trains, these vehicles leverage nuclear energy's high energy density, yet their development presents unique technical challenges and opportunities.

7.3.3.2.2 Features and safety

The primary advantage of nuclear-powered freight trucks and lorries lies in their ability for long-term autonomous operation without frequent conventional fuel replenishment. This capability is particularly suited for regions where fuel supply is difficult, such as polar and remote areas. Nuclear-powered vehicles can operate in extreme environments, reducing dependence on traditional fuels, lowering transportation costs, minimizing carbon emissions, and ensuring safe transport.

7.3.3.3 Radioisotope vehicles

7.3.3.3.1 Radioisotope vehicle concept

Radioisotope vehicles utilize RTGs as their power source. RTGs generate heat through the natural decay of radioactive isotopes like plutonium-238. This heat is then converted into electricity via the thermoelectric effect, powering the vehicle. This technology has proven successful in space probes and remote energy supply applications, offering long operational lifespans and high reliability.

7.3.3.3.2 Features and safety

The primary advantage of radioisotope vehicles lies in their extended operational range, making them particularly suitable for extreme environments or regions lacking conventional fuel supplies, such as polar expeditions or military applications. However, this technology also faces challenges related to safety and public acceptance. While RTGs meet stringent safety standards, ensuring the secure containment of isotope materials in accident scenarios remains a critical concern. In addition, the use and management of radioactive materials require rigorous regulatory oversight, adding complexity and cost to the technology.

7.3.3.4 Applications of mobile heat pipe small reactors in other fields such as sand and desert regions

7.3.3.4.1 Concept

Mobile heat pipe small reactors represent a specialized type of SMR. The term "mobile" highlights their transportability, enabling flexible deployment across diverse application needs and scenarios; "heat pipe" denotes the reactor's critical heat transfer component, whereas "small reactor" indicates its reduced scale and power output compared to traditional large nuclear power plants.

7.3.3.4.2 Applications

It can provide stable power supply to remote areas disconnected from the grid, such as deserts, mountainous regions, and polar environments, meeting local residents' domestic electricity needs and industrial production demands; supports infrastructure development, aquaculture, and desalination industries on islands, addressing energy shortages to promote economic growth and ecological conservation; provides reliable energy for military bases, naval vessels, and border outposts, enhancing operational autonomy and sustainability; supplies high-temperature steam or hot water for heat-intensive industrial processes such as chemical manufacturing, metallurgy, and food processing, meeting production requirements. In recent years, SMR technology has opened new possibilities for developing nuclear-powered trucks and lorries. SMRs' compact size, modular design, and flexible installation characteristics make them suitable for land-based transport vehicles.

7.3.3.4.3 Examples

US companies Oklo and Vertiv are exploring the integration of SMRs with electric drive systems to enhance energy efficiency and operational flexibility.

7.3.4 Challenges and development of nuclear technology in transportation

7.3.4.1 Technical reliability

Nuclear technology challenges include reactor safety, cooling system reliability, and vehicle size and weight design.

7.3.4.2 Safety assurance

Nuclear applications carry safety risks, as operations generate radioactive materials. Safety relies on multiple layers of barriers, specialized protective equipment, and monitoring systems, alongside robust accident prevention and emergency response plans.

7.3.4.3 Equipment transportation

Transportation and deployment demand stringent requirements for nuclear fuel and equipment, necessitating specialized containers and secure route planning.

7.3.4.4 Environmental facility challenges

Adaptation to complex environments is required, including foundation treatment for high-altitude sites, cold-weather equipment for polar regions, and corrosion prevention for island facilities.

7.3.4.5 Acceptance and environmental policy

Public acceptance of nuclear energy and related environmental policies are critical factors.

7.3.4.6 Safety, efficiency, and sociopolitical factors

Future efforts must advance the development of safer, more efficient nuclear propulsion systems while addressing societal and policy challenges.

7.4 Nuclear technology and military development

7.4.1 Military applications: Objectives and history

7.4.1.1 Objectives

The primary objective of applying nuclear technology in the military domain is to enhance a nation's military deterrence and combat effectiveness. Possessing nuclear weapons enables a country to occupy a pivotal position in the international political and military landscape, exerting strategic deterrence against potential adversaries and curbing their potential military adventurism. The application of nuclear propulsion technology endows combat platforms such as warships and submarines with superior mobility, endurance, and stealth capabilities, thereby elevating naval operational efficiency and strategic deployment capacity. Military radiation monitoring aims to promptly detect nuclear threats, ensure the safety of military operations in nuclear environments, and protect the physical health of military personnel.

7.4.1.2 Scope

As a vital safeguard for national security the development and maintenance of nuclear weapons remain highly classified state affairs. Although the existence of nuclear weapons maintains a degree of strategic balance internationally, their use and proliferation remain focal points of global concern. Although nuclear weapons are unlikely to disappear in the foreseeable future, the trend toward their operational deployment has slowed, with greater emphasis placed on the reliability and security of nuclear arsenals. The development and control of new nuclear weapons, along with understanding the physical principles governing nuclear weapons, remain critical issues requiring attention and resolution. In the military sphere, nuclear technology is primarily applied to nuclear weapons development, military power generation, and radiation monitoring. Nuclear weapons—including atomic bombs, hydrogen bombs, and neutron bombs—serve strategic deterrence and strike capabilities. Nuclear propulsion technology powers military vessels like warships and submarines, providing robust and sustained energy. Military radiation monitoring employs nuclear techniques to track environmental radiation levels in real time, safeguarding personnel and operational environments.

7.4.1.3 Development history

7.4.1.3.1 Inception phase

In the 1940s the United States launched the Manhattan Project, successfully developing the world's first atomic bomb—marking nuclear technology's formal entry into military applications [11]. Subsequently the United States dropped atomic bombs on Hiroshima and Nagasaki, Japan, demonstrating nuclear weapons' immense destructive power and fundamentally altering the nature of warfare and humanity's understanding of conflict.

7.4.1.3.2 Arms race phase

Following World War II the Cold War standoff between the United States and the Soviet Union ignited an intense nuclear arms race. Both sides relentlessly developed and produced various types of nuclear weapons, including hydrogen bombs and intercontinental ballistic missiles, rapidly expanding their arsenals. Concurrently, nuclear propulsion technology began military application, with the United States and former Soviet Union successively deploying nuclear-powered submarines and aircraft carriers, enhancing naval combat capabilities and strategic deterrence.

7.4.1.3.3 Restriction and control phase

Since the 1970s the international community has progressively recognized the dangers of nuclear arms races. Efforts to limit nuclear weapons development and proliferation began through a series of arms control treaties, such as the Treaty on the Non-Proliferation of Nuclear Weapons. However, competition among nations in nuclear technology research and development persists, with increased emphasis on miniaturizing and precision-targeting nuclear weapons, as well as improving and upgrading nuclear propulsion technology.

7.4.2 Nuclear weapons application

7.4.2.1 The two atomic bombs dropped by the United States on Hiroshima and Nagasaki

The advent of nuclear weapons fundamentally altered the nature of modern warfare. Toward the end of World War II, the two atomic bombs dropped by the United States on Hiroshima and Nagasaki in Japan demonstrated for the first time the immense destructive power of nuclear weapons. This not only brought the war to an end but also ushered in the nuclear age. This event not only made the world aware of the devastating power of nuclear weapons but also directly led to the realignment of the global strategic landscape after the war. With the Cold War's conclusion, nuclear weapons became a crucial pillar of strategic equilibrium among nations.

7.4.2.2 Nuclear-armed superpowers engage in arsenal competition to demonstrate deterrence

The United States and the Soviet Union engaged in an intense arms race in nuclear weapons, striving to ensure their security through larger and more powerful arsenals. Against this backdrop the theory of nuclear deterrence emerged, becoming a key tool for major powers to maintain international security.

7.4.2.3 Continuous advancements in nuclear weapons

Nuclear weapons encompass atomic bombs, hydrogen bombs, and neutron bombs. Atomic bombs harness nuclear fission reactions to release immense energy, generating powerful shockwaves, light radiation, nuclear radiation, and electromagnetic pulses to inflict devastating damage on targets. Hydrogen bombs utilize nuclear fusion reactions, yielding greater destructive power than atomic bombs. The neutron bomb is a compact nuclear weapon primarily employing high-energy neutron radiation as its lethal agent. It inflicts severe casualties on personnel while causing relatively minor damage to structures and facilities.

In 2025, the US nuclear security agency commenced production of the latest thermonuclear gravity bomb, the B61-13, with a maximum yield of approximately 360 kt—24 times the yield of the atomic bomb dropped on Hiroshima. In September 2025 the US Navy successfully test-fired four unarmed Trident II D5 ballistic missiles from an Ohio-class ballistic missile submarine in waters near Florida. In October 2025, Russia tested the nuclear-powered cruise missile "Poseidon" and successfully developed the nuclear-powered submersible "Poseidon." On September 3, 2025, during the military parade commemorating the 80th anniversary of the victory in the War of Resistance Against Japanese Aggression, China's DF-5C nuclear missile, with a strike capability covering the entire globe, demonstrated the nation's latest advancements in nuclear weapons research.

7.4.3 Military radiation monitoring

7.4.3.1 Radiation monitoring equipment and technology

Utilizing nuclear technology to develop various radiation monitoring equipment and techniques, such as radiation detectors and dosimeters, for monitoring radiation levels at nuclear facilities, military bases, and battlefield environments.

7.4.3.2 Satellite and UAV real-time monitoring

Monitoring equipment can also be deployed on platforms such as satellites and unmanned aerial vehicles (UAVs) to enable large-scale, real-time radiation monitoring. This provides environmental data support for military operations and ensures personnel safety.

7.4.4 Military nuclear power facility applications

7.4.4.1 Characteristics and value

The application of nuclear propulsion technology has significantly enhanced the endurance and strategic deployment capabilities of military platforms. This advancement is particularly evident in the development of naval forces. Nuclear submarines and nuclear-powered aircraft carriers, with their extended operational ranges and high stealth capabilities, have become core assets in the strategic naval doctrines of major powers. They played an irreplaceable role during the Cold War and in subsequent international military landscapes.

7.4.4.2 Nuclear submarines

7.4.4.2.1 Characteristics

The Cold War witnessed intense competition between the United States and the Soviet Union in nuclear submarine design and construction. The US Ohio-class and Soviet Typhoon-class submarines epitomized this era [12]. Capable of carrying multiple ballistic missiles, they executed prolonged strategic patrols worldwide, forming a vital component of national strategic nuclear deterrence. The stealth and endurance of nuclear submarines allowed them to operate within the blind spots of enemy defense systems, ensuring the reliability and surprise of nuclear strikes. Nuclear-powered submarines were the core application in this field, utilizing nuclear reactors for sustained propulsion. This enabled submarines to operate for extended periods without resupply, significantly enhancing their operational endurance.

7.4.4.2.2 United States

In recent years, significant advancements have been made in nuclear submarine technology, including efficient reactor designs and advanced fuel management systems. The US Navy's Columbia-class submarines incorporate advanced compact high-temperature gas-cooled reactors and improved uranium fuel, enhancing reliability and safety in extreme environments. In August 2025 the US President abruptly responded to remarks by former Russian President Medvedev, dispatching two nuclear submarines as a critical emergency signal. In October 2025, President Trump further announced the resumption of nuclear testing and authorized South Korea to construct nuclear submarines on US territory.

7.4.4.2.3 United Kingdom

British nuclear submarines form a critical component of its naval power, primarily comprising attack submarines and ballistic missile submarines. Attack submarines like the Astute-class represent the United Kingdom's post-Cold War design for large-scale nuclear attack submarines. The Astute-class entered service in August 2010, possessing global operational and multipurpose capabilities to execute antisubmarine warfare, antiship, and land-attack missions. It is armed with Tomahawk Block IV land-attack missiles, submarine-launched Harpoon missiles, Harpoon torpedoes, and employs pumpjet propulsion technology, delivering outstanding stealth performance.

7.4.4.2.4 Russia

Russia has achieved significant progress in nuclear submarine development, advancing the construction and commissioning of the Borei-A class strategic nuclear submarines and the Borei-M class attack multipurpose nuclear submarines. The Arkhangelsk (Borei-M class) entered service at the end of 2024; the Grand Duke Pozharsky (Borei-A class) was formally commissioned in July 2025. Regarding modernization: Older nuclear submarines, such as the Project 971 and Project 949AM classes, are undergoing upgrades. For instance the Project 971M Leopard's modernization is nearing completion, whereas the Project 949AM Irkutsk has completed upgrades enabling it to carry Kalibr, Oniks, and Zircon missiles. In addition, Russia has strengthened nuclear submarine deployments in the Far East. For instance the Alexander III and Krasnoyarsk have joined the Pacific Fleet's submarine force. In the future the Belgorod and Khabarovsk nuclear submarines will also arrive in the Far East, carrying Poseidon nuclear torpedoes.

7.4.4.2.5 China

China's nuclear-powered submarines are currently undergoing continuous development and refinement. Although China's first-generation attack submarines suffered from issues such as high underwater noise levels, they laid the foundation for subsequent advancements. China's second-generation strategic submarines represent the nation's first strategic nuclear submarines capable of reliable underwater second-strike nuclear retaliation.

7.4.4.3 Nuclear-powered aircraft carriers

7.4.4.3.1 Characteristics

Nuclear-powered aircraft carriers represent another significant military application of nuclear technology. Their nuclear propulsion systems enable prolonged combat readiness in theater without frequent resupply, substantially enhancing operational efficiency. These carriers utilize nuclear reactors to support rapid global deployment and extended maritime operations. Nuclear-powered carriers, such as the US Navy's Nimitz-class and Ford-class, are equipped with advanced nuclear reactor systems, substantially increasing their endurance, operational range, and service life. Design and technological advancements enable these carriers to execute diverse tactical missions, including global strike and maritime control, significantly bolstering naval strategic strike capabilities.

7.4.4.3.2 United States

The development of US nuclear-powered carriers began in the 1950s, culminating in the commissioning of the world's first nuclear-powered carrier, the USS Enterprise, in 1961. Starting in the 1970s the Nimitz-class nuclear-powered carriers were progressively built and commissioned, with a total of 10 vessels becoming the backbone of the US Navy. Entering the 21st century, the United States introduced the more advanced Ford-class nuclear-powered carrier, with the lead ship, USS Ford, commissioned in 2017. In 2025 the second ship of the Ford class, USS John F. Kennedy, completed its first nuclear power test and is scheduled for delivery in March 2027.

7.4.4.3.3 United Kingdom

The United Kingdom has pioneered numerous innovations in carrier technology, such as angled decks and arresting wires. Under its CVF program for next-generation carriers the United Kingdom ultimately built two Queen Elizabeth-class carriers equipped with gas turbine propulsion systems, with the lead ship commissioned in 2017.

7.4.4.3.4 Russia

As of 2025, its only active aircraft carrier, the Kuznetsov, is facing the fate of being scrapped or sold because of multiple accidents and high maintenance costs. The development strategy of the Russian Navy has shifted toward nuclear submarines as the core, abandoning traditional aircraft carrier formations.

7.4.4.3.5 China

The construction of China's first nuclear aircraft carrier is rapidly advancing. The most remarkable breakthrough of this aircraft carrier lies in its nuclear power system. Analysis suggests that it may adopt advanced fourth-generation molten salt reactor technology, which has significant advantages in fuel efficiency and endurance compared to traditional PWRs. The adoption of nuclear power will give aircraft carriers almost unlimited endurance, enabling them to carry out missions in distant waters for a long time and providing ample power for their electromagnetic catapult systems, laser weapons that may be equipped in the future, and more.

7.4.4.4 Nuclear-powered vessels

7.4.4.4.1 Characteristics

Compared to conventionally powered ships, nuclear-powered vessels offer advantages such as greater power output, extended endurance, and virtually unlimited operational range. This enables large ships to carry more supplies, travel faster and farther, and serve a wider range of purposes, making them suitable for long-range combat operations and extended patrol missions. Currently deployed primarily in military applications, they form a crucial component of naval forces worldwide, significantly enhancing a nation's maritime military capabilities and strategic deterrence.

7.4.4.4.2 United States

Under the US Navy's "Future 30-Year Shipbuilding Plan," 364 vessels will be procured over the next three decades. America's nuclear-powered ship development centers on nuclear-powered aircraft carriers and submarines. Nuclear propulsion systems provide near-limitless endurance and strategic deterrence, sustaining US military superiority across global waters.

7.4.4.4.3 United Kingdom

The UK Ministry of Defence conducted an information-gathering exercise to explore the feasibility of using fourth-generation nuclear energy systems to power large surface vessels, with a particular focus on SMRs.

7.4.4.4.4 Russia

On August 18, 2025 the nuclear-powered cruiser Admiral Nakhimov, following extensive modernization, commenced sea trials. The upgraded vessel is equipped with new naval guns, a universal vertical launch system, and the capability to deploy various weapons, including Zircon hypersonic missiles, significantly enhancing the Russian Navy's long-range strike capabilities.

7.4.5 Military energy supply in remote areas

7.4.5.1 Solutions for addressing military energy supply challenges in remote regions

Remote areas such as high mountains, polar regions, deserts, and islands face significant difficulties in traditional energy supply because of harsh geography and climate. Land transportation is constrained by terrain, resulting in high costs and low efficiency; air transport has limited capacity and is prohibitively expensive; maritime transport also proves inconvenient for remote islands. Fuel storage is prone to degradation from environmental factors, leading to unstable power supply. Nuclear energy, however, offers high energy density, enabling substantial power generation from minimal fuel quantities, thereby reducing transportation and storage burdens. Its extended operational cycles provide long-term stable power supply, eliminating the need for frequent resupply and ensuring reliable energy support for sustained military base operations and combat missions.

7.4.5.2 Nuclear energy technologies suitable for remote areas

SMRs employ modular designs, factory-prefabricated and assembled on-site, enabling short construction cycles and low costs. With capacities ranging from 10 to 300 MW, their compact structures suit space-constrained remote locations, whereas integrated designs enhance safety. For example, the US NuScale reactor delivers 50 MW per module and incorporates passive safety systems, making it suitable for powering island bases and desalination. Nuclear microreactors operate below 1 MW, are compact for easy transport, and can be integrated into shipping containers. Utilizing advanced technologies, they achieve high operational efficiency. Models like the Aurora nuclear microreactor can meet energy demands at air bases in extreme cold environments.

7.4.6 Trends in military application development

7.4.6.1 Nuclear weapons

Nuclear weapons may evolve toward greater miniaturization, precision, and intelligence, enhancing strike accuracy, reducing collateral damage, and improving penetration capabilities and survivability. Simultaneously the development of new nuclear weapons may become a focus for some nations, such as nuclear electromagnetic pulse bombs and nuclear bunker-busting bombs, to meet diverse operational requirements. The US "Z Machine" employs powerful electromagnetic pulse compression technology to generate extreme temperatures and pressures, simulating the physical environment at the core of a nuclear explosion. France's "Laser Mégajoule" facility is an inertial confinement laser fusion experiment platform serving both fundamental physics research and the simulation of processes related to nuclear fusion warheads. The Advanced Light Source (ALS) at Lawrence Berkeley National Laboratory in the United States utilizes some of its experimental platforms to support nonexplosive testing of nuclear materials for the National Nuclear Security Administration (NNSA). Similar high-energy X-ray experimental capabilities are also available at facilities such as the French Synchrotron Radiation Center and Russia's Kurchatov Institute.

7.4.6.2 Nuclear propulsion

Further enhancing the safety, reliability, and efficiency of nuclear propulsion systems, developing compact, high-performance reactors to enable nuclear power applications across a broader range of military platforms—including small vessels, UAVs, and even land combat platforms. In addition, exploring novel nuclear propulsion technologies such as nuclear rocket engines to power future space-based military operations.

7.4.6.3 Military radiation monitoring

Advancements in quantum technology, artificial intelligence, and new materials will render military radiation monitoring systems more intelligent, precise, and compact. Technologies, such as quantum sensors and smart monitoring networks, will see widespread adoption, enabling faster and more accurate detection of trace radiation levels. This facilitates early warning and precise assessment of nuclear threats.

7.4.6.4 Deep space and deep sea applications

Technological advancements may extend the use of military nuclear power facilities to deep space and deep-sea platforms. For instance the concept of nuclear-powered spacecraft is evolving, enabling long-term space cruising for strategic surveillance and strike missions. Deep-sea operations will also benefit from nuclear propulsion technology. Future nuclear-powered submersibles will be capable of executing complex military missions in extreme marine environments, such as reconnaissance, attacking enemy undersea facilities, or safeguarding national maritime interests. The development and deployment of these new nuclear-powered systems will significantly expand the operational reach and tactical options of military forces, enabling more flexible and effective strategic deployments for nations.

7.5 Nuclear technology and medical development

7.5.1 Diagnosis of common human diseases

7.5.1.1 Target scope

Fundamental human diseases refer to those chronic conditions that persistently impact human health. These diseases may be hereditary or acquired, affecting multiple bodily systems. Nuclear technology finds extensive application in medicine, demonstrating unique advantages and value, particularly in addressing fundamental human diseases. This includes imaging diagnostics for tumors, coronary heart disease, and kidney diseases, as well as the treatment of hyperthyroidism [13]. The demand for nuclear technology during medical emergencies is detailed in Section 7.6.3.

7.5.1.2 Various imaging technologies

7.5.1.2.1 SPECT imaging

Single-photon emission computed tomography (SPECT) imaging utilizes single-energy gamma photons emitted during radionuclide decay. The most commonly used nuclide is technetium-99m (99mTc), which emits low-energy gamma photons at 140 keV. SPECT is applicable for diverse examinations, including renal dynamic imaging, whole-body bone imaging, cerebral perfusion imaging, myocardial perfusion imaging, thyroid imaging, lymphatic imaging, and gastrointestinal bleeding detection, providing critical diagnostic information for conditions such as tumors.

7.5.1.2.2 PET imaging

PET (positron emission tomography) imaging utilizes radioactive isotopes that emit positrons, such as fluorine-18 (^{18}F) and carbon-11 (^{11}C). When positrons interact with electrons within the body, they annihilate, producing high-energy gamma photons. Detectors capture these photons to generate images. PET excels in detecting malignant tumors because of its sensitivity, accuracy, specificity, and precise localization. The examination is brief (approximately 10 minutes), convenient, and highly efficient. PET identifies abnormal cellular metabolism within lesions, detecting abnormalities earlier than conventional computed tomography (CT) scans. Its success rate in identifying malignant tumors exceeds 90%, earning it the reputation as the "X-ray vision for detecting malignant tumors."

7.5.1.2.3 MRI imaging

MRI (magnetic resonance imaging) contrast agents alter tissue signal intensity, with gadolinium-based agents being the most widely used positive contrast agents [14]. Molecular imaging techniques have also advanced, utilizing labeling of endogenous specific molecules as endogenous contrast agents. Different

MRI sequences play crucial roles in diagnosing and treating diseases at various stages, such as diffusion-weighted imaging (DWI) and T2-weighted fluid-attenuated inversion recovery (FLAIR) sequences. MRS technology reflects material changes in brain cells, whereas MRI angiography enhances the diagnostic rate of aneurysms and cerebral arteriovenous malformations. Amino acid proton transfer imaging reflects tissue pH changes, enabling lesion visualization and detection of dynamic alterations. 5.0 T MRI technology surpasses conventional MRI in imaging speed, resolution, detection of small lesions, and diagnostic accuracy. MR microscopy achieves spatial resolution as high as 4 μm, facilitating animal model research for diseases and drug development.

7.5.1.2.4 Imaging diagnosis of coronary heart disease

Radionuclide myocardial perfusion imaging is a noninvasive, highly accurate diagnostic method that uses trace amounts of radioactive isotopes to evaluate coronary blood flow perfusion, myocardial cell function, and cell viability. It directly visualizes myocardial blood flow perfusion and functional status, enabling definitive determination of myocardial ischemia or infarction, thereby providing crucial diagnostic evidence for coronary heart disease.

7.5.1.2.5 Imaging diagnosis of kidney diseases

Renal imaging employs radionuclide tracing technology to visualize kidneys, evaluating renal function and related conditions such as urinary tract obstruction or renovascular hypertension [15]. Renal imaging is categorized into dynamic and static renal imaging, differing based on examination objectives, methodologies, and imaging agents used. As a noninvasive, safe, simple, and cost-effective imaging technique, a single examination provides comprehensive clinical information reflecting renal blood flow, function, morphology, and urinary tract patency. It remains an indispensable method for diagnosing urinary system disorders.

7.5.2 Human health examination

7.5.2.1 X-ray examination

7.5.2.1.1 Definition

X-ray examination is a diagnostic method using X-rays to detect diseases. It is categorized into general examination, specialized examination, and contrast examination. Fluoroscopy is a simple and commonly used examination method that allows observation of organ morphology and functional changes from different angles.

7.5.2.1.2 Method

X-ray examination primarily utilizes the penetrating power, differential absorption, photosensitivity, and fluorescence effects of X-rays. Tissues of varying densities absorb X-rays to different degrees, creating shadows of differing densities on the fluoroscope screen or X-ray film.

7.5.2.1.3 Applications

X-ray examination is widely used for assessing the gastrointestinal tract, skeletal system, chest, and genitourinary system.

7.5.2.1.4 Advantages

It provides clear, detailed, and straightforward imaging, making it one of the most fundamental medical imaging techniques currently available.

7.5.2.2 CT (computed tomography)

7.5.2.2.1 Definition

CT employs precisely collimated X-ray beams, gamma rays, or ultrasound waves in conjunction with highly sensitive detectors to perform cross-sectional scans around specific body regions.

7.5.2.2.2 Method

Detectors capture X-rays transmitted through a specific tissue layer. These are converted into visible light, then transformed into electrical signals via photoelectric conversion. The signals are processed by a computer to reconstruct the image of that layer. CT technology has evolved from first to fifth generation, achieving faster scanning speeds and higher image quality. Modern systems like 64-slice CT can acquire high-quality multislice images in minimal time, making them particularly suitable for cardiovascular angiography and acute trauma patients.

7.5.2.2.3 Applications

CT examinations are widely used across medical fields.

7.5.2.2.4 Advantages

It can detect minute lesions in any body region, holding significant diagnostic value for tumors, cerebrovascular diseases, and trauma.

7.5.2.3 Magnetic resonance imaging (MRI)

7.5.2.3.1 Definition

MRI utilizes signals generated by hydrogen nuclei within the human body when exposed to a magnetic field. These signals are processed by a computer to reconstruct images. Hydrogen protons are abundant and widely distributed throughout the body. When placed in a strong magnetic field, hydrogen nuclei align and move according to the field's direction. Upon application of radiofrequency pulses, hydrogen nuclei release energy signals. These signals are received and processed by a computer to generate images.

7.5.2.3.2 Method

MRI operates on the principle of nuclear magnetic resonance. The body is first placed within a strong magnetic field, aligning hydrogen nuclei (protons) along the field direction. Specific-frequency radiofrequency pulses are then applied, causing protons to absorb energy, resonate, and deviate from the magnetic field direction. When the RF pulse ceases, the protons release energy and return to their original state. At this point, detectors capture the electromagnetic wave signals emitted by the protons. After computer processing, based on differences in proton density and relaxation times (T1, T2) within various tissues, cross-sectional images of the body's internal structures are reconstructed, enabling noninvasive imaging diagnosis of human tissues.

7.5.2.3.3 Applications

MRI examinations offer advantages such as multiparameter imaging, high soft tissue resolution, and absence of ionizing radiation. It clearly displays internal structures, particularly soft tissues, providing unique diagnostic capabilities for conditions affecting the brain, spinal cord, joints, and soft tissues. MRI is widely applied across clinical medicine, including examinations of the nervous system, cardiovascular system, and musculoskeletal system. Simultaneously, with continuous technological advancement, MRI has achieved significant progress in functional imaging and molecular imaging.

7.5.2.3.4 Advantages

MRI produces exceptionally clear and detailed images, greatly enhancing diagnostic efficiency and avoiding invasive procedures like exploratory thoracotomy or laparotomy [16]. As MRI does not utilize harmful X-rays or contrast agents prone to allergic reactions, it poses no harm to the human body. MRI enables multiangle, multiplane imaging of various body regions with high resolution, objectively and precisely displaying anatomical structures and their spatial relationships. This facilitates more accurate localization and characterization of lesions. It holds significant value in diagnosing diseases across all body systems, particularly in the early detection of tumors.

7.5.3 Treatment of major human diseases

7.5.3.1 Radiation therapy for tumors

7.5.3.1.1 Etiology analysis

A tumor is a neoplasm formed when a single cell in a localized tissue loses normal genetic regulation of its growth under the influence of various carcinogenic factors, leading to clonal abnormal proliferation. Tumor cells are generally considered monoclonal, meaning all tumor cells within a single tumor are descendants of a single mutated cell.

7.5.3.1.2 Treatment methods

Radiation therapy represents a significant application of nuclear technology in medicine and serves as a vital treatment for malignant tumors. It employs radiation devices or sources to deliver X-rays, gamma rays, beta rays, and other forms of radiation to the tumor site, thereby destroying cancer cells.

7.5.3.1.3 Treatment types

Radiation therapy is primarily categorized into two types: external beam therapy and brachytherapy. External beam therapy equipment includes medical electron accelerators, gamma knives, and proton/heavy ion therapy systems. Brachytherapy primarily refers to afterloaders and similar devices.

7.5.3.1.4 External beam therapy

7.5.3.1.4.1 Indications This approach is suitable for patients with tumors that are inoperable, carry a risk of residual disease or recurrence after surgery, or are unresponsive to chemotherapy. In addition, patients with distant metastases from cancers like lung or liver cancer—where local treatment is insufficient—often utilize nuclear technology for external beam therapy to suppress tumor growth and alleviate symptoms.

7.5.3.1.4.2 Process External beam therapy delivers high-energy radiation precisely to the tumor site from outside the body, serving as a core component of comprehensive cancer treatment.

7.5.3.1.4.3 Efficacy Its technological evolution, deeply integrated with nuclear physics, imaging technology, and computer science, continues to push the boundaries of tumor treatment. Medical linear accelerators, for instance, generate high-energy X-rays or electron beams suitable for treating deep-seated tumors. Gamma Knife utilizes gamma rays from cobalt-60 to deliver a single high-dose, focused irradiation for precise treatment. Proton and heavy ion therapy is internationally recognized as a cutting-edge radiotherapy technique. It delivers potent irradiation to tumor lesions while sparing surrounding healthy tissues, maximizing therapeutic efficacy.

7.5.3.1.5 Brachytherapy

7.5.3.1.5.1 Indications Primarily used for treating cervical cancer, prostate cancer, breast cancer, and skin cancers. It involves the close-range irradiation of tumors using radiation emitted from radioactive nuclides.

7.5.3.1.5.2 Description Brachytherapy is a crucial modality within nuclear technology-based radiation therapy. Unlike conventional external beam radiation therapy, it involves placing the radiation source directly within or adjacent to the tumor tissue for irradiation, also known as "internal irradiation."

7.5.3.1.5.3 Efficacy The primary advantage of this approach lies in delivering high radiation doses to the tumor tissue while significantly reducing radiation exposure to surrounding healthy tissues. This enhances treatment efficacy while minimizing radiation-induced side effects.

7.5.3.2 Treatment of hyperthyroidism

7.5.3.2.1 Target population

Hyperthyroidism symptoms are often associated with increased metabolic activity, affecting multiple body systems. Common symptoms include palpitations and rapid heartbeat, even at rest; excessive sweating and heat intolerance, with moist skin; increased appetite but weight loss, sometimes accompanied by more frequent bowel movements; emotional irritability, anxiety, difficulty concentrating, and possible insomnia; muscle weakness and hand tremors, especially noticeable when holding objects; possible thyroid enlargement and neck thickening; some individuals may experience protruding eyeballs, dry eyes, photophobia, and blurred vision; women may have menstrual irregularities such as reduced flow and prolonged cycles, while men occasionally develop breast tissue.

7.5.3.2.2 Content

Diagnosing hyperthyroidism requires integrating symptoms, thyroid function tests, and imaging assessments. First, consider typical symptoms such as palpitations, excessive sweating, and weight loss. The core diagnostic step involves blood tests to assess thyroid function: checking for decreased thyroid-stimulating hormone (TSH) levels and elevated free triiodothyronine (FT3) and free thyroxine (FT4) levels. Thyroid-related antibodies, such as TSH receptor antibodies (TRAb), are also tested to help determine the cause. In some cases a thyroid ultrasound is performed to examine thyroid morphology and blood flow, ruling out other thyroid disorders. A minority may require further thyroid scintigraphy to clarify thyroid function status. Diagnosis is confirmed by integrating all these findings.

7.5.3.2.3 Treatment outcomes

Hyperthyroidism treatment primarily involves three approaches, selected based on disease severity: first, antithyroid medication suppresses thyroid hormone synthesis. This prolonged regimen (typically 1.5–2 years) requires regular thyroid function monitoring to adjust dosage, suitable for mild-to-moderate hyperthyroidism or younger patients. Second, radioactive iodine therapy involves oral administration of radioactive iodine to destroy thyroid tissue and reduce hormone secretion. Although highly effective, it may cause hypothyroidism, necessitating long-term hormone replacement therapy. Third, surgical treatment involves partial or total thyroidectomy, suitable for cases with significant thyroid enlargement, compression symptoms, or poor response to medication/radioactive iodine therapy. Hormone replacement therapy may also be required postsurgery. Treatment outcomes for hyperthyroidism vary by modality but are generally manageable. Medication requires long-term adherence, with potential recurrence in some cases. Radioactive iodine and surgery offer reliable efficacy, effectively controlling hyperthyroidism in most patients. However, the latter two methods carry a higher risk of hypothyroidism, necessitating lifelong thyroid hormone replacement. With standardized treatment, normal daily functioning is typically unaffected.

7.5.3.3 Treatment of bone metastases

7.5.3.3.1 Indications

Bone metastases represent a common complication of malignant tumors, often presenting with symptoms such as intractable pain and restricted mobility. Patients frequently experience progressively worsening bone pain, initially manifesting as intermittent dull aches that evolve into persistent, severe pain as the disease advances. This pain intensifies at night or during activity, directly linked to tumor cells invading the periosteum, destroying bone tissue, and releasing pain-inducing substances. Some patients develop pathological fractures because of weakened bone strength. Spinal compression fractures may cause height loss, kyphosis, or even spinal cord compression, leading to limb numbness and muscle weakness. Rib fractures manifest as severe pain during coughing or deep breathing.

7.5.3.3.2 Content

Leveraging the enhanced osteoblastic activity at tumor bone metastases, which presents as "hot spots" with radioactive concentration, this method enables whole-body skeletal screening in a single session. It is suitable for initial screening and monitoring treatment efficacy. Strontium-89, as a radiopharmaceutical in nuclear medicine, has demonstrated significant efficacy in treating bone metastases. Strontium-89 adheres to metastatic sites, emitting beta rays for internal radiation therapy that destroys metastatic tumor cells while alleviating bone pain and inhibiting lesion growth. It is indicated for patients with multiple bone metastases, unresectable primary bone tumors or residual disease postsurgery, and malignancies prone to bone metastasis (e.g., prostate cancer, breast cancer).

7.5.3.3.3 Efficacy

Nuclear technology demonstrates verifiable efficacy in treating bone metastases. By emitting beta rays that target and accumulate at bone metastasis sites, radionuclides such as strontium-89 and samarium-153 effectively suppress tumor cell activity and alleviate bone pain, with pain relief rates exceeding 70% in some patients. It also reduces osteoclast activity and slows bone destruction progression, significantly improving quality of life for patients with multiple bone metastases. This approach is particularly suitable for those unresponsive to conventional pain management, though potential adverse effects like bone marrow suppression require monitoring. Comprehensive assessment of the patient's physical condition and lesion distribution is essential prior to treatment.

7.5.3.4 Treatment of prostate cancer

7.5.3.4.1 Target population

Prostate cancer is the most common malignant tumor of the male reproductive system, with incidence increasing with age. Its incidence shows marked regional variation, being higher in Europe and America. The etiology of prostate cancer remains unclear, potentially involving genetic, environmental, and sex hormone factors. Approximately 98% of prostate cancers are adenocarcinomas, often originating in the peripheral zone of the atrophied prostate and frequently presenting as multiple lesions.

7.5.3.4.2 Content

Nuclear technology has established a precise molecular imaging system for prostate cancer diagnosis. Targeted specific tracers enable highly sensitive tumor identification and localization [17]. By integrating metabolic information with anatomical structures, this approach provides critical evidence for clinical staging, treatment selection (e.g., radical surgery, radiotherapy target delineation), and efficacy monitoring, significantly enhancing diagnostic accuracy and treatment precision.

7.5.3.4.3 Therapeutic efficacy

This approach utilizes high-energy particles emitted by radioactive isotopes to destroy cancer cells. Its advantages include noninvasiveness, high precision, and short treatment duration. It precisely targets cancer cells while minimizing damage to surrounding healthy tissues.

7.6 Nuclear technology and high-tech development

7.6.1 Development of advanced reactor designs

7.6.1.1 Key advanced technologies used in fourth-generation nuclear reactors

7.6.1.1.1 Advanced coolant technologies

Fourth-generation nuclear reactors employ various advanced coolants to enhance performance and safety. High-temperature gas-cooled reactors utilize helium, whose excellent thermal conductivity and chemical stability support safe operation at elevated temperatures, thereby improving thermal efficiency. Liquid metal–cooled reactors employ liquid sodium, potassium, lithium, lead, or lead-bismuth alloys, with emerging research exploring materials such as gallium and bismuth. These coolants feature low melting points, high boiling points, and excellent thermal conductivity, enabling efficient operation at low pressure and high temperature. Molten salt reactors utilize molten salts, whose high thermal conductivity and boiling points ensure stable operation at high temperatures and low pressures. The liquid state of molten salts also optimizes fuel handling and recycling processes.

7.6.1.1.2 Fast reactor technology

Approximately half of fourth-generation reactors employ fast neutron reactor technology. This approach utilizes fast neutrons for fission, enabling efficient utilization of uranium-238 (^{238}U) and thorium-232 (^{232}Th), thereby enhancing fuel efficiency. Fast reactors possess strong breeding capabilities, generating additional fissile material to mitigate nuclear fuel resource scarcity. They also significantly reduce high-level radioactive waste, enhance safety, and support closed fuel cycles, improving nuclear energy sustainability.

7.6.1.1.3 Advanced safety systems

Advanced safety systems are critical for fourth-generation reactors. Passive safety systems automatically cool the reactor during power outages without external intervention; real-time monitoring and intelligent control systems instantly detect and adjust anomalies; redundancy design and fault-tolerant mechanisms enhance system reliability. These measures ensure overall safety, reduce accident risks, strengthen public trust, and promote the sustainable development of nuclear energy.

7.6.1.1.4 Efficient energy conversion technology

High-efficiency energy conversion technology is a defining feature of fourth-generation reactors. By optimizing thermoelectric conversion efficiency, it enhances energy utilization while reducing fuel requirements. This lowers costs and environmental impact, bolstering reactor economics and competitiveness [18].

7.6.1.1.5 Heat pipe technology

7.6.1.1.5.1 Technical principle Heat pipe reactor cooling technology utilizes heat pipes as core heat exchange components. Its fundamental principle involves the phase-change cycle of the working fluid within the heat pipe: liquid working fluid absorbs heat and vaporizes in the core evaporation section; gaseous working fluid releases heat and liquefies in the condensation section; gravity or capillary forces then return the fluid, passively transferring core heat away.

7.6.1.1.5.2 Technical features This technology offers significant advantages: the system eliminates the need for traditional primary loop equipment, features a compact structure for easy transport, and is suitable for scenarios such as power supply in remote areas and deep-sea exploration. Heat pipes provide passive, independent cooling; failure of a single pipe does not affect the overall system, ensuring high safety. Typical components include fuel rods, heat pipes, control devices, and reflectors. Heat can be utilized through thermoelectric conversion or propulsion systems.

7.6.1.1.5.3 Examples The United States has demonstrated this technology for space nuclear power systems, whereas China is advancing novel design research. However, overcoming phase transition challenges during startup remains a key technical hurdle for high-temperature metal heat pipes. Improper power control can lead to dryout or startup failure. Led by Tsinghua University with participation from Southeast University, the national key R&D project "Megawatt-Class Intelligent Modular Metal Diffusive Heat Pipe Reactor Mobile Nuclear Power System" investigates heat pipe-cooled reactors and has produced a prototype model of a mobile vehicle-mounted heat pipe reactor. Professor Zhou Tao's team at Southeast University has applied for and obtained the patent "A two-dimensional heat pipe and heat exchanger," proposing a novel heat pipe and heat exchanger design.

7.6.1.2 Key advanced technologies used in small modular reactors

7.6.1.2.1 Modular design

The modular design of SMRs enables reactor components to be prefabricated and assembled in modules at factories, thereby shortening construction cycles and reducing costs. In addition, this design offers flexible scalability, allowing modules to be added or removed based on actual demand to meet varying power requirements. It also simplifies on-site installation and maintenance while enhancing safety and reliability.

7.6.1.2.2 Compact reactor design

Compact reactor design optimizes space utilization, enabling operation within a smaller footprint. This approach not only reduces construction and maintenance costs but also simplifies on-site installation, enhancing overall reactor economics and flexibility.

7.6.1.2.3 Advanced fuel technologies

7.6.1.2.3.1 Fundamental structure High-density fuel and enhanced fuel cycles extend operational lifetimes while minimizing waste generation. These technologies optimize fuel management and resource utilization, bolstering reactor sustainability. Novel annular fuel elements feature a unique construction comprising dual cladding layers and annular core blocks.

7.6.1.2.3.2 Technical characteristics This structure enables simultaneous cooling from both the inner and outer surfaces of the element. Compared to conventional solid fuel rods, it offers significant advantages in heat transfer. On one hand, it reduces the thickness of the radial heat transfer path within the core, significantly lowering peak temperatures inside the core. On the other hand, it increases the wetted perimeter of the coolant, substantially enhancing cooling capacity.

7.6.1.2.3.3 Example Research at the Massachusetts Institute of Technology (MIT) demonstrates that applying this fuel in PWRs can increase power density by up to 150% while maintaining or even enhancing safety margins.

7.6.1.2.4 Intelligent monitoring and control systems

Advanced intelligent monitoring and control technologies enable operators to monitor reactor status in real time and perform remote operations. This includes advanced sensor networks and data analysis tools that enhance reactor operational efficiency and safety.

7.6.1.3 Key advanced technologies used in fusion reactors

7.6.1.3.1 Superconducting magnet technology

Superconducting magnet technology is a core technology in nuclear fusion research. It utilizes high-temperature superconducting (HTS) materials, such as yttrium barium copper oxide (YBCO) and bismuth strontium calcium copper oxide (BSCCO), to generate powerful magnetic fields that effectively confine high-temperature plasma. This technology is crucial for maintaining plasma stability and achieving sustained fusion reactions.

7.6.1.3.2 Plasma control and diagnostics technology

Plasma control technologies, including radiofrequency heating and neutral beam injection (NBI), are employed to heat and stabilize the plasma. Advanced diagnostic techniques, such as laser scattering and X-ray imaging, enable real-time monitoring of plasma conditions, ensuring optimal operation to maintain reactor safety and efficiency.

7.6.1.3.3 Fusion fuel cycle technology

Fusion fuel cycle technology primarily involves the production and management of tritium. Tritium is generated through neutron reactions with lithium target materials and must be managed via tritium extraction and storage systems. These technologies ensure the efficient utilization of tritium fuel and support the long-term stable operation of reactors [19].

7.6.2 Advances in nuclear physics

7.6.2.1 Contributions of nuclear technology to particle physics research

7.6.2.1.1 High-energy accelerator technology

High-energy accelerators are central to particle physics research, generating new particles by accelerating particles to near-light speeds and colliding them. In recent years, nuclear technology has propelled advancements in this field. The upgraded Large Hadron Collider (LHC) increased collision energies to 13.6 TeV in 2022, delivering more precise data that confirmed the Higgs boson and laid the groundwork for exploring dark matter and supersymmetric particles.

7.6.2.2 Advanced detector technology

Advanced detector technology is crucial for particle physics experiments, particularly high-energy collision studies. Modern detectors, such as silicon detectors and scintillator detectors, leverage nuclear technology principles to efficiently capture particle data.

The LHC, Compact Muon Solenoid (CMS), and Atlas detectors have significantly enhanced experimental precision and efficiency through novel materials and designs, such as high-resolution silicon detectors. These technologies have enhanced particle detection capabilities, enabling scientists to more accurately measure particle energy and trajectories, thereby gaining deeper insights into minute interactions between particles.

7.6.2.3 Contributions of nuclear technology to nuclear physics research

7.6.2.3.1 High-energy nuclear experiment facilities

7.6.2.3.1.1 Technical features High-energy nuclear experiment facilities serve as core tools in nuclear physics research. By accelerating particles to extremely high energies through nuclear technology, they enable the study of particle interactions with atomic nuclei.

7.6.2.3.1.2 Examples The International Heavy Ion Collider (FAIR) project officially commenced on November 7, 2007. Located at the GSI Helmholtz Centre for Heavy Ion Research in Darmstadt, Germany, it is scheduled for completion in 2027. Upon completion, it will generate matter and antimatter, enabling cutting-edge research in nuclear physics and related fields. Its aim is to provide higher energy heavy ion beams to explore the properties of hadronic matter. FAIR's design objectives include deepening understanding of nuclear matter's fundamental structure through high-energy nuclear collision experiments and investigating nuclear physics phenomena under extreme conditions.

The China Spallation Neutron Source (CSNS), a major national scientific infrastructure, is situated in Songshan Lake Science City, Dongguan. Serving as a powerful platform for fundamental scientific research, it plays an irreplaceable role in fields critical to national strategic needs, such as novel energy materials and high-performance alloys. It also attracts global researchers for collaboration, positioning Dongguan as an international center for cutting-edge nuclear technology applications and multidisciplinary research.

7.6.2.3.2 Precision measurement technology

7.6.2.3.2.1 Technical features Precision measurement technology is crucial in nuclear physics research. High-precision measurement equipment and techniques, such as precision mass spectrometers and high-sensitivity detection systems, enable accurate determination of atomic nuclei's mass, lifetime, and decay characteristics. These technologies not only allow scientists to verify nuclear physics theories but also aid in studying nuclear forces and their mechanisms.

7.6.2.3.2.2 Examples Research utilizing high-precision mass spectrometry has successfully determined the masses of superheavy elements, advancing our understanding of new elements and nuclear structure [20].

7.6.2.4 Contributions of nuclear technology to astrophysical research

7.6.2.4.1 Study of stellar internal nuclear reactions

7.6.2.4.1.1 Technical features Nuclear technology is indispensable for investigating nuclear reactions within stars. Stellar energy originates from nuclear fusion processes such as the proton–proton chain and the carbon–nitrogen–oxygen (CNO) cycle.

7.6.2.4.1.2 Examples China's Jinping Underground Nuclear Astrophysics Experiment (JUNA) directly measured the reaction cross section for fluorine radiation capture—a key nuclear astrophysical reaction—extending measurements to the world's lowest energy range. It discovered a new resonance at 225 keV, explaining the origin of calcium abundance in the universe's oldest known stars.

7.6.2.4.2 Neutrino astronomy

7.6.2.4.2.1 Technical features Neutrinos, significant byproducts of nuclear reactions, can penetrate celestial bodies and Earth's crust to reveal internal information. Advances in nuclear technology have enhanced neutrino detection capabilities.

7.6.2.4.2.2 Examples The Super-Kamiokande detector at the Hosokawa Mine in Kamioka Town, Hida City, Gifu Prefecture, Japan, and the IceCube

Neutrino Observatory near the Amundsen-Scott South Pole Station in Antarctica have detected neutrinos from supernovae and other astrophysical events. These detections have confirmed neutrino properties and revealed mechanisms of supernova explosions and the distribution of interstellar matter.

7.6.2.4.3 Gamma-ray astronomy

7.6.2.4.3.1 Technical features Gamma-ray astronomy also benefits from advances in nuclear technology. Gamma rays are the highest energy photons in the universe, produced by violent nuclear reactions and high-energy particle acceleration processes.

7.6.2.4.3.2 Examples Modern gamma-ray detectors, such as the Fermi Gamma-ray Space Telescope, utilize nuclear technology to detect and analyze these high-energy rays, revealing the nature and distribution of gamma-ray sources. These techniques are crucial for understanding extreme celestial phenomena such as black holes, pulsars, and gamma-ray bursts.

7.6.2.5 Contributions of nuclear technology to dark matter detection research

7.6.2.5.1 Development of high-sensitivity detectors

7.6.2.5.1.1 Technical features In the field of dark matter detection the development of high-sensitivity detectors is crucial. Nuclear technology provides core support for the design of these detectors.

7.6.2.5.1.2 Examples The liquid xenon detector XENON1T is one of the most advanced devices. It leverages nuclear technology principles to enhance detection sensitivity. By optimizing signals generated from electron-ion pairs in xenon atoms, it achieves exceptionally high sensitivity to dark matter candidate particles, resulting in significant breakthroughs.

7.6.2.5.2 Low background noise technology

7.6.2.5.2.1 Technical features To effectively detect dark matter signals, interference from environmental background noise must be overcome. Nuclear technology plays a vital role in reducing detector background noise.

7.6.2.5.2.2 Example The LUX-ZEPLIN (LZ) detector at the Sanford Underground Research Facility in South Dakota, United States, significantly reduces background noise by being constructed deep underground and utilizing ultrahigh-purity liquid xenon. This design enhances detection precision, enabling the LZ detector to perform more detailed analysis of dark matter particles and further advancing dark matter detection technology.

7.6.3 Development of new materials technology

7.6.3.1 Research and development of high-temperature resistant material technology

7.6.3.1.1 High-temperature simulation testing material technology

Nuclear technology has promoted the research of high-temperature materials by simulating extreme conditions inside nuclear reactors to test and improve the performance of materials. For example, carbide materials and tungsten-based alloys have become candidate materials for high-temperature components in nuclear reactors because of their excellent melting points and high-temperature stability. The melting point of carbides (such as silicon carbide) exceeds 2500°C, while the melting point of tungsten-based alloys is as high as 3422°C. These characteristics enable these materials to operate stably for a long time in high-temperature environments.

7.6.3.1.2 High-temperature synthesis and rapid cooling technology

The development of nuclear technology has promoted the application of advanced manufacturing technologies, such as high-temperature synthesis and rapid cooling technology. These technologies can produce materials with excellent high-temperature performance. For example, through improved chemical vapor deposition (CVD) and physical vapor deposition (PVD) processes, researchers can generate high-temperature stable protective coatings on the surfaces of tungsten-based alloys and other high-temperature resistant materials, thereby enhancing their high-temperature oxidation resistance and corrosion resistance.

7.6.3.1.3 Synchrotron radiation and neutron diffraction techniques

Nuclear technology has also promoted a deeper understanding of material properties, including thermal conductivity, coefficient of thermal expansion, and chemical stability. By using advanced experimental equipment such as synchrotron radiation and neutron diffraction techniques, researchers can observe the structural changes of materials at the microscopic level, thereby better predicting their behavior in high-temperature environments. These research results provide a scientific basis for the design and optimization of materials.

7.6.3.2 Research and development of radiation-resistant materials

7.6.3.2.1 Developing new radiation-resistant materials

The advancement of nuclear technology has promoted the development of new radiation-resistant materials. The development of radiation-resistant materials involves the design of new alloys and composite materials that can effectively mitigate radiation-induced damage. In recent years, research has focused on the development of high-entropy alloys and carbide materials, which have attracted attention because of their unique atomic structure and high radiation stability.

High-entropy alloys increase their radiation resistance and stability by randomizing their constituent elements. Carbide materials such as silicon carbide have become potential materials for future nuclear reactors because of their excellent high-temperature and radiation stability [21].

7.6.3.2.2 Promote research on the mechanism of radiation damage

Nuclear technology has also promoted the study of radiation damage mechanisms, including radiation-induced microstructural changes and their impact on material properties. By using advanced characterization techniques such as transmission electron microscopy and atomic force microscopy, scientists are able to observe the microscopic changes in materials during irradiation, thereby improving material design and enhancing their radiation resistance. These research results provide important scientific basis for the development of more durable nuclear materials.

7.6.3.3 Research and development of superconducting materials

7.6.3.3.1 Evaluating superconducting material performance and stability

High-energy accelerators and nuclear technologies provide the high-energy particle beams essential for superconducting material research. These beams simulate extreme radiation environments, aiding in the evaluation of superconducting materials' stability and performance under real-world conditions. Through neutron irradiation experiments, researchers can investigate structural changes and performance degradation in materials under intense radiation, thereby optimizing material design. This is crucial for developing radiation-resistant superconductors suitable for nuclear fusion and high-energy physics experiments.

7.6.3.3.2 Enhancing the performance and stability of high-temperature superconductors

Nuclear technology drives the development and optimization of high-temperature superconductors. Research employing nuclear techniques has improved the performance of nitride and perovskite superconductors, increasing their critical temperatures and stability under strong magnetic fields. These advances provide more reliable material support for superconducting magnets in fusion reactors and particle accelerators.

7.6.3.4 Research and development of nanomaterials

7.6.3.4.1 Preparation and modification of nanomaterials

Nuclear technology provides high-energy particle beams and radiation sources, which are crucial for the preparation and modification of nanomaterials. For example, neutron sources and electron beams can be used for radiation modification of materials, altering the structure and properties of nanomaterials.

Through radiation treatment, researchers can precisely control the particle size, morphology, as well as the chemical and physical properties of nanomaterials. This method is particularly suitable for synthesizing nanoparticles with specific functions, such as catalysts and drug carriers.

7.6.3.4.2 Characterization and analysis of nanomaterials

Nuclear technology is used to characterize and analyze nanomaterials. Advanced nuclear technologies such as neutron scattering and synchrotron X-ray imaging can provide structural information of nanomaterials, including atomic-level arrangement and defects. These technologies help scientists understand the behavior of nanomaterials under various environmental conditions, thereby guiding material design and optimization. For example, neutron scattering technology can reveal the magnetic, elastic, and dynamic properties of nanomaterials, which is crucial for developing high-performance nanomaterials [25].

7.6.4 Research and development of biomimetic materials

7.6.4.1 High-performance coolant materials with bionic structures

In the field of nuclear power generation the combination of nuclear and biomimetic materials is expected to enhance the safety and efficiency of nuclear power plants. Develop high-performance coolant materials with biomimetic structures, such as mimicking the efficient heat dissipation mechanism of blood flow in biological blood vessels, and design new coolant channel structures. Improve the heat transfer efficiency of the coolant, better dissipate the heat generated by the reactor, reduce the core temperature, and minimize the risk of accidents.

7.6.4.2 Bionic self-healing materials

Bionic self-healing materials can be used to manufacture critical components for nuclear power plants, such as pipes and valves. These materials can promptly repair damage caused by corrosion, fatigue, and other factors during operation, extending equipment lifespan and ensuring stable plant operation.

7.6.4.3 Energy storage and conversion bionic materials

In renewable energy and battery sectors, bionic materials offer advantages for energy storage and conversion. Combined with nuclear technology for material performance optimization, they can be used to develop new energy storage devices such as batteries and supercapacitors, as well as high-efficiency energy conversion devices such as solar cells and fuel cells. This drives the energy sector toward greater efficiency and sustainability.

7.7 Nuclear technology and healthy living development

7.7.1 Common scenarios of nuclear radiation exposure

7.7.1.1 Radiation levels near nuclear power plants

7.7.1.1.1 Radiation sources and standards

Residents near nuclear power plants are exposed to radiation from natural background sources and trace amounts of radioactive substances released into the environment during plant operations. The state has established strict emission limits for radioactive materials to ensure radiation safety for the environment and public near nuclear facilities. Current national standards stipulate that the annual additional radiation dose from all sources (excluding background radiation and medical exposure) must be less than 1 mSv. For nuclear power plants the annual public dose resulting from radioactive emissions must be less than 0.25 mSv.

7.7.1.1.2 Actual radiation levels and quantities

The maximum public dose caused by annual radioactive emissions from nuclear power plants operating in China is about a few microsieverts, far below the national standard limit. Compared to natural background radiation exposure (approximately 2.4 mSv/year), the impact of nuclear power plants on the environment and the public is very small in magnitude.

7.7.1.1.3 Risk and safety

Multiple domestic and international studies confirm that nuclear facilities like power plants pose no impact to nearby residents under normal operating conditions. For example, a 2004 Chinese survey of residents near the Qinshan Nuclear Power Plant after 11 years of operation showed radiation doses within natural background levels, with no significant difference in malignant tumor mortality rates before and after plant operation.

7.7.1.2 Radiation exposure in medical examinations

7.7.1.2.1 Types of radiological examinations

Medical radiation constitutes one of the primary sources of additional radiation exposure for individuals [26]. However, the doses involved are far below thresholds capable of causing direct harm to the human body (e.g., 100 mSv). Under normal circumstances, radiation doses from medical examinations are safe for humans. Theoretically, they pose no harm unless individuals undergo frequent radiological examinations. Medical procedures involving radiation primarily include X-ray examinations, CT scans, and nuclear medicine tests (e.g., PET–CT, SPECT).

7.7.1.2.2 Radiation doses

7.7.1.2.2.1 X-ray examinations Single-exposure doses range from approximately 0.01 to 0.7 mSv; chest radiography exposes patients to 0.1 to 1 mSv; barium meal examinations can reach 2 to 7 mSv.

7.7.1.2.2.2 CT scans Radiation dose varies based on the scanned area and number of scans, typically ranging from several mSv to tens of mSv. Contrast-enhanced CT scans may involve higher doses.

7.7.1.2.2.3 PET–CT scans Radiation exposure is relatively high, with a single scan exposing the patient to approximately 15 mSv.

7.7.2 Applications of nuclear technology in daily life

7.7.2.1 Nuclear power generation

7.7.2.1.1 Concept

Nuclear power generation involves utilizing nuclear fuel (such as uranium) to undergo controlled nuclear fission within a nuclear reactor. The released energy heats water into high-temperature, high-pressure steam, which drives a turbine to rotate and power a generator. Its core principle is the stable release of heat through a chain reaction, achieving power supply via the conversion sequence: "nuclear energy → thermal energy → mechanical energy → electrical energy." For detailed content, see Section 7.1.2.

7.7.2.1.2 Characteristics

Nuclear power plants comprise nuclear reactors, steam generation systems, and other components. They offer advantages including high energy density (1 kg of uranium fission energy equals 2700 t of coal), low carbon emissions, and stable power supply. See Section 7.1.2 for details.

7.7.2.1.3 Example

Nuclear power generation converts energy through nuclear reactors. Taking China's Qinshan Nuclear Power Plant as an example, its core is a PWR that releases thermal energy through controlled nuclear fission. This heats primary loop water to high temperature and pressure, which then transfers heat to the secondary loop water via heat exchangers. This process generates high-temperature steam that drives turbine rotation, thereby powering generators to produce electricity. Finally the electricity is transmitted to the grid to meet industrial production and residential electricity demands. By 2025 the Qinshan Nuclear Power Plant had operated safely for over 30 years, accumulating more than 700 billion kW-hours of electricity generation. This equates to reducing standard coal consumption by approximately 220 million t and cutting carbon dioxide emissions by about 580 million t, exemplifying a clean and stable energy supply.

7.7.2.2 *Nuclear heating*

7.7.2.2.1 Concept

The reactor within a nuclear-powered industrial park serves as the core facility, playing a vital role in the park's operations and related sector development. It utilizes nuclear fission or fusion reactions to generate thermal energy, which is then converted into electricity or other forms of energy. Nuclear heating harnesses the thermal energy produced during nuclear power generation to provide heating services for specific users or even entire industrial parks. See Section 7.1.3 for details.

7.7.2.2.2 Characteristics

This approach is not only clean and environmentally friendly but also highly efficient. The nuclear heating process involves extracting steam from the nuclear power unit, which undergoes multiple heating stages through heat exchange stations before transferring heat to the heating system. This method ensures a secure isolation between the nuclear power plant and users, preventing radioactive substances from entering the heating pipelines. Compared to traditional heating methods, nuclear heating offers significant advantages. First, nuclear energy is a clean energy source that produces negligible emissions of pollutants such as carbon dioxide and sulfur dioxide, effectively reducing air pollution and contributing to carbon reduction goals. Second, nuclear heating offers high stability, unaffected by climate or seasonal factors, providing users with continuous and reliable heating services. In addition, nuclear heating boasts high thermal efficiency, effectively reducing energy consumption and improving energy utilization efficiency.

7.7.2.2.3 Example

In Haiyan, Zhejiang, a nuclear heating demonstration project has been successfully implemented, providing local residents with safe, zero-carbon, and economical heating services. Jiangsu Xuyi's nuclear heating system employs advanced PWR technology. Heat generated by the nuclear reactor warms primary loop water, which then transfers heat to secondary loop water, producing high-temperature, high-pressure steam. This steam drives turbines for electricity generation, whereas residual heat is transferred via heat exchangers to circulating water in the heating network, ultimately delivering warmth to users.

7.7.2.3 *Nuclear hydrogen production*

7.7.2.3.1 Concept

Nuclear hydrogen production refers to the process of using nuclear energy as a clean, stable energy source to decompose water into hydrogen and oxygen through specific technologies. Its core lies in utilizing thermal or electrical energy from nuclear reactors to drive reactions such as water electrolysis and thermochemical cycles. This enables the efficient production of high-purity hydrogen without consuming fossil fuels or emitting carbon. The resulting hydrogen can be applied in fuel cells, industrial feedstocks, energy storage, and other fields, serving as a key technological pathway to bridge the nuclear and hydrogen energy industries and advance "zero-carbon" goals.

7.7.2.3.2 Characteristics

Nuclear hydrogen production exhibits distinct advantages in cleanliness, stability, and efficiency. Utilizing nuclear energy as its source the entire process consumes no fossil fuels and generates negligible carbon emissions, aligning with zero-carbon development requirements. Nuclear reactors provide continuous 24-hour energy supply, unaffected by seasonal or weather conditions, ensuring stable thermal or electrical energy for hydrogen production and maintaining operational continuity. Furthermore, nuclear energy can directly utilize reactor high temperatures through technologies such as high-temperature steam electrolysis and thermochemical cycles, achieving higher energy conversion efficiency than traditional fossil fuel-based hydrogen production. Leveraging existing nuclear power infrastructure reduces the cost and technical barriers for large-scale hydrogen production, making it a vital bridge between nuclear and hydrogen energy that underpins future energy transitions.

7.7.2.3.3 Examples

Tsinghua University has achieved continuous, stable hydrogen production through iodine-sulfur cycle and high-temperature electrolysis experiments. China's Tianwan Nuclear Power hydrogen production project employs PE water electrolysis technology, accumulating over 30,000 m^3 of hydrogen by August 2025 with 90% electrolysis efficiency and 99.999% hydrogen purity.

7.7.2.4 Mobile nuclear power unit

7.7.2.4.1 Concept

The mobile nuclear power unit is an ultracompact, transportable advanced nuclear energy system utilizing liquid metal cooling. It typically employs fourth-generation lead-based fast neutron reactors with specialized liquid metal as the coolant.

7.7.2.4.2 Features

The mobile nuclear power unit boasts ultrasafety, ultracompact size, ultralong endurance, and high flexibility. Its lead alloy coolant enables operation at atmospheric pressure, eliminating high-pressure hazards and effectively preventing nuclear accidents. Its container-sized dimensions facilitate transportation via trucks, providing energy supply across maritime, land, air, and space applications. Through ingenious design, it fully leverages the breeding properties of ^{238}U in nuclear fuel, enabling a single fuel load to last over 10 years. In addition, it offers multiple voltage outputs for flexible deployment, providing stable power to remote areas, islands, mining sites, and more. It can also be used in industrial production, seawater desalination, and other fields, making it an efficient, clean, and flexible energy supply device.

7.7.2.4.3 Examples

The micronuclear power pack can serve as a space power source, supporting deep-space exploration and manned lunar missions. The plus version can function as a ship propulsion system, powering submarines and large vessels. Furthermore the nuclear power pack holds significant application potential in isotope production, seawater desalination, and other fields.

7.7.2.5 Delicious foods

7.7.2.5.1 Concept

A typical application of nuclear technology in the field of delicious foods is irradiation preservation technology.

7.7.2.5.2 Characteristics

The International Atomic Energy Agency (IAEA) and other organizations confirm that irradiated foods meeting standards contain no radioactive residues. Their safety is comparable to preservation methods like heating or refrigeration, preserving ingredients' flavor while enhancing food supply chain security.

7.7.2.5.3 Examples

This technology uses gamma rays emitted by radioactive isotopes such as cobalt-60 or high-energy electron beams generated by electron accelerators to irradiate food. The radiation penetrates food surfaces, disrupting the DNA structure of microorganisms (such as bacteria and mold), inhibiting insect egg hatching and sprouting in fruits and vegetables, thereby extending shelf life without significantly altering color, texture, or nutritional content. For example, irradiated fruits reduce mold growth; meat eliminates parasites; spices and dried goods lower microbial contamination risks.

7.7.2.6 Health and medical care

7.7.2.6.1 Concept

Nuclear technology applications in health and medical care primarily rely on radioactive isotopes and nuclear techniques, as detailed in Sections 7.6.3 and 7.7.5.

7.7.2.6.2 Characteristics

These applications integrate nuclear physics with medicine, demonstrating unique value in disease diagnosis, treatment, and life support. While involving radiation, strict regulations ensure safety, making them crucial technical pillars of modern healthcare.

7.7.2.6.3 Examples

Gamma rays from cobalt-60 enable precise radiation therapy for cancer by targeting and destroying tumor cell DNA. Iodine-131 is used for diagnosing and treating thyroid diseases, where its beta rays selectively eliminate diseased cells. Nuclear medicine imaging like PET–CT employs fluorine-18 isotope-labeled drugs to track metabolic activity and locate early-stage lesions. Furthermore, radioactive isotopes enable the creation of nuclear batteries, providing long-term power for implantable devices like pacemakers.

7.7.2.7 Environmental protection

7.7.2.7.1 Concept

Nuclear technology plays a multidimensional role in environmental protection.

7.7.2.7.2 Characteristics

As a tool, nuclear technology offers innovative solutions for carbon reduction, pollution control, and resource recycling.

7.7.2.7.3 Examples

On one hand, nuclear power generation serves as a low-carbon energy source. The energy released from the fission of 1 kg of uranium is equivalent to that from 2700 t of standard coal. A single megawatt-class nuclear power plant reduces annual carbon dioxide emissions by approximately 6 million t, significantly mitigating the greenhouse effect and atmospheric pollution caused by fossil fuels. On the other hand, nuclear technology can be applied to pollutant treatment. For instance, electron beam irradiation technology treats industrial wastewater by using high-energy electrons to disrupt the molecular structure of organic compounds in water, decomposing them into harmless substances. Radioisotope tracer technology precisely tracks pollutant migration pathways, providing data support for soil and groundwater remediation. In addition, nuclear irradiation can process solid waste such as scrap tires and plastics, enhancing material recycling rates through radiation modification and contributing to the development of a "waste-free society."

7.7.3 Impact of nuclear wastewater discharge into the ocean

7.7.3.1 Impact on marine ecology

7.7.3.1.1 Destruction of marine ecosystems

High-dose radiation directly causes the death of marine organisms. In addition, prolonged exposure to low-dose radiation induces genetic mutations in marine life, leading to alterations in body structure, morphology, growth, and reproduction. Such mutations may inflict long-term damage on marine ecosystems. The accumulation of radioactive substances within marine organisms can also propagate through the food chain to higher tier organisms, further exacerbating ecosystem disruption.

7.7.3.1.2 Mortality and mutation of marine organisms

Nuclear wastewater contains radioactive substances such as cesium, cobalt, strontium, uranium, and plutonium. These substances enter marine ecosystems via water currents, causing radiation contamination of marine plants, plankton, benthic organisms, and others. Radioactive materials damage the genetic material of these organisms, leading to abnormal growth, reduced reproductive capacity, or even death. The disruption of ecosystems further destabilizes the entire food chain, potentially reducing marine biodiversity and upsetting ecological balance.

7.7.3.1.3 Contamination of coastal areas

Radioactive substances in the wastewater can drift with ocean currents to shorelines, contaminating coastal zones and surrounding areas. This causes long-term impacts on coastal ecosystems and shoreline habitats, damaging local environments and residents' living conditions.

7.7.3.2 Impact on food safety

The discharge of nuclear wastewater raises public concerns about the safety of seafood. Consumers may reduce or avoid purchasing seafood because of safety worries.

7.7.3.3 Impact on residents

7.7.3.3.1 Health risks

Residents may be exposed to radioactive substances through consumption of contaminated seafood (e.g., fish, shellfish) or ingestion of polluted seawater. Long-term exposure increases risks of health issues, including cancer, genetic mutations, and immune system damage. Residents near discharge sites face particularly heightened health risks.

7.7.3.3.2 Sociopsychological and attitudinal impacts

The discharge of nuclear wastewater may alter public perceptions and attitudes toward nuclear energy. Once the wastewater enters the ocean and demonstrates tangible impacts, public skepticism and resistance regarding the safety and sustainability of nuclear power could intensify. Such sociopsychological shifts may further impede the development and utilization of nuclear energy.

7.7.3.4 Economic impacts

Fishermen may reduce or cease fishing activities because of concerns about seafood contamination, negatively impacting local fisheries economies. Beyond fisheries the discharge could also adversely affect coastal tourism and related industries. Public apprehension about radioactive substances may lead to decreased tourist activity or avoidance of contaminated areas.

7.7.3.5 Addressing challenges

Advanced nuclear technology development has made the probability of severe accidents at nuclear power plants extremely low. Nevertheless, many countries have undertaken targeted national research projects [22]. Emergency measures and response strategies are also in place for the unlikely event of a major nuclear accident. Currently the Zhou Tao team at Southeast University is leading the development of the standard "Technical Specifications for Identification and Measurement of Radioactive Isotopes in the Ocean" (T/CIET1673-2025), providing standardized guidance for nuclide identification and measurement in the event of a nuclear power plant accident.

7.8 Nuclear technology and human future development

7.8.1 Fields of nuclear technology and human survival and development

7.8.1.1 Energy sector

Nuclear technology offers a clean, efficient energy source capable of meeting humanity's energy demands while reducing reliance on fossil fuels. Nuclear power generates substantial electricity for society, lowers power costs, minimizes environmental pollution, and serves as a low-carbon energy solution. Artificial intelligence powered by nuclear energy analyzes and forecasts energy system data through machine learning and deep learning technologies. This optimizes every stage of energy production, transmission, storage, and consumption, enabling efficient utilization, intelligent dispatch, and low-carbon transformation. It propels the energy sector toward intelligent and sustainable development. Nuclear energy is widely applied in energy production, providing humanity with a reliable and efficient energy source. Nuclear power generation can meet growing energy demands while reducing reliance on fossil fuels. Compared to traditional energy sources, nuclear power generation reduces greenhouse gas emissions, playing a positive role in addressing climate change. In addition, nuclear energy has relatively low operating costs and plays a crucial role in ensuring stable energy supply.

7.8.1.2 Medical field

Nuclear technology is widely applied in disease diagnosis and treatment, such as nuclear medicine and radiation therapy. Nuclear medicine relies on radioactive tracers to achieve precise localization and functional assessment of lesions. By injecting tracers containing specific radionuclides into the body, researchers leverage the nuclides' decay properties and targeted accumulation capabilities. Combined with imaging techniques like PET, this approach enables dynamic tracking of tracer distribution and metabolic processes within the body. Radiation therapy, centered on the biological effects of ionizing radiation, precisely controls radiation dose and coverage to disrupt tumor cell DNA structures and inhibit proliferation. It has become a vital component in comprehensive treatments for malignant tumors such as lung cancer, breast cancer, and cervical cancer.

7.8.1.3 Environmental protection

Nuclear technology also makes significant contributions to environmental protection. Compared to traditional energy sources, nuclear power generates fewer greenhouse gas emissions during energy production, reducing air pollution and mitigating climate change impacts. In addition, nuclear power generation avoids the accumulation of large volumes of coal gangue and tailings, thereby decreasing soil and water contamination. Although nuclear power generation carries inherent safety risks, appropriate safety measures and technological advancements can minimize accident probabilities and ensure secure operation. Nuclear wastewater treatment employs techniques such as ion exchange, evaporation concentration, and membrane separation to remove radionuclides. For instance, Japan's Fukushima nuclear wastewater is treated using the ALPS system, though tritium remains difficult to fully eliminate. Treated wastewater undergoes rigorous testing and must meet standards before discharge to minimize risks to the environment and human health.

7.8.1.4 Radiation processing applications

Nuclear technology finds applications in isotope labeling, nondestructive testing, and radiation processing. Isotope labeling tracks chemical reaction processes, nondestructive testing detects internal defects in objects, and radiation processing enhances materials' chemical and physical properties.

7.8.1.5 Scientific research

Nuclear technology expands research frontiers in physics, chemistry, and biology, solving long-standing scientific challenges.

7.8.1.6 Deep space and deep-sea exploration

Nuclear energy powers space and deep-sea missions, providing propulsion systems and satellite communications.

7.8.2 Nuclear technology and human future development

7.8.2.1 Future development goals

7.8.2.1.1 Survival and development goals

Nuclear technology holds significant importance for human survival and development. However, its advancement must prioritize safety risks and proceed under appropriate regulation and management. Therefore while promoting nuclear technology development, emphasis should be placed on safety and sustainability to achieve human survival, development, and future prosperity.

7.8.2.1.2 Energy supply goals

In the energy sector, nuclear power generation provides reliable, efficient, and environment-friendly energy, reducing dependence on fossil fuels.

7.8.2.1.3 Medical objectives

In healthcare, nuclear technology enhances diagnostic accuracy and enables effective treatment for certain diseases.

7.8.2.1.4 Environmental protection objectives

Regarding environmental conservation, nuclear power generation reduces greenhouse gas emissions and minimizes pollution of soil and water resources.

7.8.2.2 Advantages and limitations

7.8.2.2.1 Minimal environmental impact

As a clean energy source, nuclear technology exerts significantly less environmental impact than fossil fuels. Nuclear power generates far fewer carbon dioxide emissions than traditional fossil energy sources. In addition, nuclear power plants can produce substantial electricity, reducing greenhouse gas emissions into the atmosphere.

7.8.2.2.2 Severe nuclear accidents pose major hazards

The environmental and health impacts of a severe nuclear accident can be catastrophic. Even a severe nuclear accident with an extremely low probability of occurrence can, if mishandled, release large amounts of radioactive material, contaminating the surrounding soil and water sources and posing a serious threat to the health and safety of people living nearby.

7.8.2.2.3 The issue of nuclear waste storage

The storage of nuclear waste after accidents is also a serious problem, as long-term storage and disposal require significant costs and technology. Nuclear energy has a significant impact and role on the future of humanity. However, nuclear energy also needs to continuously advance technological progress and improve safety to ensure future safety and sustainable development.

7.8.2.3 Challenges and development

7.8.2.3.1 Addressing nuclear accident risks

Management and control of nuclear accident risks require further refinement. By enhancing safety measures and equipment, establishing stringent standards, and fostering a robust safety culture, we can reduce accident occurrence and prevent harm to the environment and humanity.

7.8.2.3.2 Addressing nuclear waste treatment and storage challenges

Nuclear waste treatment and storage demand safer and more rational approaches, necessitating the exploration of sustainable and environment-friendly disposal methods.

7.8.2.3.3 Meeting innovation and development challenges

Advancements in nuclear technology bring considerable hope and opportunities for humanity's future. However, achieving sustainable development in nuclear technology demands overcoming numerous difficulties and challenges. Continuous innovation is essential to explore core scientific principles, develop new technologies and equipment, and enhance efficiency and safety. Simultaneously the nuclear technology application industry constitutes a vital component of new productive forces. In November 2025, China released its first "Action Plan for Strengthening the Nuclear Technology Application Industry Chain," actively advancing the high-quality leap forward of the nuclear technology sector.

7.8.3 Building a community with a shared future for nuclear safety

7.8.3.1 A community of shared future for nuclear security must emphasize symbiotic security

7.8.3.1.1 The interdependent community is an objective reality

As world history unfolds and global interactions deepen, the world increasingly becomes an interdependent community where "we are inseparable from one another." Our survival depends on the existence of others. General Secretary Xi Jinping emphasized, "The impact of nuclear security incidents transcends national borders. In an era of interconnectedness, no country can tackle such challenges alone, nor can any country remain unaffected."

7.8.3.1.2 Subjects and objects maintain a symbiotic relationship

The advancement of nuclear technology has placed modern society within a web of risks. Issues such as nuclear arms races, nuclear proliferation, nuclear terrorism, and nuclear leaks threaten human values and dignity. Humanity must strengthen symbiotic relationships to address nuclear security challenges and uphold security. Symbiosis establishes the ontological security of the self; symbiosis and security represent the internal and external dimensions of harmonious coexistence with others. The nuclear security community of shared destiny bridges the limitations of subject–object dualistic thinking, transforming the relationship between subject and object into a unified symbiotic whole.

7.8.3.1.3 Subject and object unified in practical development

The nuclear security community of shared destiny recognizes that the current alienated state of interactions in nuclear security stems from the distorted development of human nuclear practices. The relational dynamics inherent in this community guide subjects to undertake virtuous planning through correct actions within complex contexts. This approach adheres to the objective laws of international economic and political systems while consistently treating humanity as the purpose of development, unifying subject and object within practical engagement.

7.8.3.1.4 The nuclear security community of shared destiny requires active construction

The nuclear security community of shared destiny does not treat the objects of nuclear practice as a world of certainty but rather grasps the uncertainty of nuclear risks from the realm of possibility. The process of nuclear practice is one of moving from possibility to reality. Therefore constructing a certain nuclear security community of shared destiny can better address the uncertainty of nuclear security risks.

7.8.3.2 The nuclear security community of shared destiny promotes shared security

7.8.3.2.1 Shared responsibility is essential in the nuclear age

In the nuclear age, humanity wields immense material power, making us citizens of a "global village" who must share responsibility, coexist peacefully, and cooperate mutually. Otherwise, it would be difficult to harness nuclear power with its enormous energy and risks. General Secretary Xi Jinping emphasized, "We should coordinate and integrate global nuclear security resources with the IAEA at the core, and leverage its expertise to serve all countries."

7.8.3.2.2 Narrow national centrism as the root of international nuclear security issues

The emergence of international nuclear security issues stems from narrow national centrism, which exacerbates the structural contradiction of unequal distribution of nuclear technology rights. This makes it difficult for actors to resist the negative impacts of immense nuclear material power, leading them into extreme self-centeredness. Within the international community, nations pursue not only the maximization of behavioral interests but also other interests in political diplomacy, military security, and cultural exchange. If actors pursue the maximization of economic interests through narrow state-centered approaches, other interests will be compromised, ultimately hindering the realization of their own interests.

7.8.3.2.3 Embracing the values of a community with a shared future for mankind is essential for sharing nuclear security

The nuclear security community with a shared future upholds a correct view of justice and interests, transcending narrow state-centered approaches. This community emphasizes balancing individual and collective interests, prioritizing the shared vision of a community with a shared future for mankind as the fundamental value perspective for mutual nuclear security. It sets sustainable human development and a better life as its core objectives. Even amid today's complex and volatile international nuclear security landscape, it advocates consciously assuming global responsibility, providing value guidance for humanity to share nuclear security in the nuclear age.

7.8.3.3 The nuclear safety community of shared destiny advocates for jointly building security

7.8.3.3.1 Jointly building security is the consensus and norm of security ethics

The nuclear safety community of shared destiny is not an abstract concept; it is an open system continuously evolving through practice. It emphasizes equality, mutual benefit, and equal rights among all participants, building a shared human consensus on nuclear safety through an open and equitable approach. Therefore jointly building security means forming a consensus and norms of security ethics through the negotiated integration of individual and collective interests.

7.8.3.3.2 The nuclear security community of shared destiny requires jointly establishing sound consultation mechanisms

In the process of jointly building the nuclear security community of shared destiny, countries establish sound consultation mechanisms, laying the foundation for consensus on nuclear security ethics. Although China first proposed the concept of the nuclear security community of shared future, China is not the dominant force in its construction but rather an active participant, promoter, and contributor. Building this community is not about establishing a new nuclear order where one center replaces another, but rather about forming an interactive relationship among multiple centers.

7.8.3.3.3 Security coconstruction is an egalitarian process of decoercion

Building security together is a process of decoercion. It opposes the use of force to impose will, advocating instead for dialogue-based equality of discourse and consultation among diverse value systems and cultural traditions. Constructing a nuclear security community of shared destiny requires equal, interactive relationships among actors. It necessitates building these relationships through collaborative security efforts, thereby reducing conflicts of will among actors and diminishing the likelihood of nuclear security issues [23].

7.8.3.4 Responsibilities in building a nuclear security community of shared destiny

7.8.3.4.1 Building a community of interdependence, reciprocity, and interaction

International nuclear security issues fundamentally reveal the inherent flaws in state-to-state relations and the drawbacks of narrow state-centered approaches. A nuclear security community of shared destiny signifies the dissolution of the subject–object dualistic mindset. It is the unity of a community of interdependence, reciprocity, and interaction built upon dialogue and consultation.

7.8.3.4.2 Autonomous choice by nations; ensuring nuclear safety is each nation's responsibility

The nuclear safety community of shared destiny resolves interest disputes through dialogue and consultation guided by a correct view of justice and interests, transcending narrow state-centered perspectives [24]. As General Secretary Xi Jinping stated, developing nuclear energy is an autonomous choice for each nation, and ensuring nuclear safety is each nation's responsibility.

7.8.3.4.3 Cobuilding a nuclear safety order based on symbiosis and shared benefits to proactively mitigate risks

Nuclear safety represents a universal aspiration for all peoples and constitutes a fundamental responsibility for nuclear-weapon states. Nuclear-weapon states should take the lead in building a community with a shared future for nuclear safety, fostering a nuclear safety order characterized by symbiosis, shared benefits, and collaborative construction, while proactively and effectively mitigating nuclear safety risks.

References

[1] L.J. Dang, X.L. Li, W.R. Zhang, et al., Design and implementation of DCS System MTS for nuclear power plant based on plug-in technology, International Symposium on Software Reliability, Industrial Safety, Cyber Security and Physical Protection for Nuclear Power Plant, Springer, Singapore, 2024, doi:10.1007/978-981-97-7051-9_4.

[2] Y.Q. Chen, B.H. Yan, The technology of shielding design for nuclear reactor: a review, Prog. Nucl. Energy 161 (2023) 1–12, doi:10.1016/j.pnucene.2023.104741.
[3] AREVA GroupNuclear Fuel Cycle Technology White Paper, AREVA Group, Paris, 2019.
[4] H.W. Liu, Economic benefits of nuclear energy in industrial applications, J. Nucl. Energy Technol. 43 (2) (2021) 32–45.
[5] D. Houssin, T. Dujardin, R. Cameron, Technology Roadmap: Nuclear Energy, International Energy Agency, Paris, 2015.
[6] D. Dong, Z. Wang, J. Guan, et al., Research on key technology and application progress of rescue robot in nuclear accident emergency situation, Nucl. Eng. Technol. 57 (6) (2025), doi:10.1016/j.net.2025.103457.
[7] Y.S. Park, S. Park, J. Kim, et al., Development of human-in-the-loop experiment system to extract evacuation behavioral features: a case of evacuees in nuclear emergencies, Nucl. Eng. Technol. (2023), doi:10.1016/j.net.2023.02.032.
[8] L. Zhang, X.M. Wang, Current status of nuclear heating technology and applications, Energy Res. Inf. 37 (3) (2022) 67–74.
[9] D.S. Wisnubroto, G.R. Sunaryo, Y.S.B. Susilo, Indonesia's RDE Program: a multifaceted approach to nuclear energy development encompassing human resource building, public acceptance, and technological innovation, Nucl. Eng. Technol. 57 (8) (2025), doi:10.1016/j.net.2025.103560.
[10] P.D. Porras, J. Sanz, L. Cizelj, W. Ambrosini, G. Pavel, Sustaining regional nuclear human capacity building in Europe, Nucl. Eng. Des. 419 (5) (2024) 112916.
[11] D. Holloway, Stalin and the Bomb: The Soviet Union and Atomic Energy 1939-1956, Yale University Press, New Haven, 1995.
[12] N. Polmar, Cold War Submarines: The Design and Construction of U.S. and Soviet Submarines, 1945-2001[M], Potomac Books, Dulles, 2004.
[13] M. Jeong, D. Han, P. Bhetariya, et al., NF2 is essential for human endoderm development, Adv. Sci. 12 (17) (2025) e2410909, doi:10.1002/advs.202410909.
[14] B.L. Lee, M. Rout, M.D.S. Wishart, Automated identification and quantification of metabolites in human fecal extracts by nuclear magnetic resonance spectroscopy, Magn. Reson. Chem. 61 (12) (2023) 705–717, doi:10.1002/mrc.5372.
[15] A. Schumacher, T.Q. Nguyen, R. Broekhuizen, et al., Structural development of the human fetal kidney: new stages and cellular dynamics in nephrogenesis, Biorxiv (2023), doi:10.1101/2023.03.24.534074.
[16] Y. Ono, H. Shirasawa, K. Takahashi, et al., O-069 The first mitotic spindle instability impacts nucleation, fragmentation, and development in human embryos, Hum. Reprod. 39 (Sup1) (2024) 1, doi:10.1093/humrep/deae108.075.
[17] H. Hou, Y. Pan, Y. Wang, et al., Development and first-in-human study of PSMA-targeted PET tracers with improved pharmacokinetic properties, Eur. J. Nucl. Med. Mol. Imaging 51 (9) (2024), doi:10.1007/s00259-024-06726-6.
[18] T. Miller, Advances in Nuclear Power Plant Efficiency, Nuclear Energy Institute, Washington, DC, 2020.
[19] M. Aydin, M. Aydin, The causal effects of nuclear fusion reactors, human development, and economic growth on nuclear energy consumption in the United States, Int. J. Energy Res. (2024) 2024, doi:10.1155/2024/2836877.
[20] L. Yongchao, L. Tong, X. Kai, et al., Research on sensor data optimization technology for thermal hydraulic experiment of nuclear reactor, Nucl. Eng. Des. (2024) 423, doi:10.1016/j.nucengdes.2024.113176.

[21] A. Ketharam, D. Pletser, N. Evans, et al., Field-assisted processing of Ce–zirconolite ceramics for nuclear applications, Int. J. Appl. Ceram. Technol. (2025), doi:10.1111/ijac.15119.
[22] Y.C. Chang, X. Zhao, Addressing Japan's disposal of nuclear-contaminated water from the perspective of international human rights law, Chin. J. Popul. Resour. Environ. 22 (1) (2024) 1–9, doi:10.1016/j.cjpre.2024.03.001.
[23] Y. Li, M.X. Liu, H. Peng, et al., Research and application of endogenous security active defense technology for domestic nuclear safety-level gateway equipment, International Symposium on Software Reliability, Industrial Safety, Cyber Security and Physical Protection for Nuclear Power Plant, Springer, Singapore, 2023, doi:10.1007/978-981-99-3455-3_20.
[24] Y. Okui, C. Cyuta, H. Yoshioka, et al., Issues of the nuclear disaster core facility through nuclear disaster training, Prehosp. Disaster Med. (2023) 38, doi:10.1017/S1049023ž23004806.
[25] G.C.A. Ferreira, P.D. Santos, M.S. Zangirolami, et al., Comparative analysis of the fatty acid composition in human, cow and goat milk and their interpositional distribution in triacylglycerols, Quim. Nova 48 (3) (2024) 1–6.
[26] A. Domingo-Muelas, R.M. Skory, A.A. Moverley, et al., Human embryo live imaging reveals nuclear DNA shedding during blastocyst expansion and biopsy, Cell (15) (2023) 186, doi:10.1097/01.ogx.0000993700.13780.f9.

Chapter 8

The future of human and nuclear civilization

Peng Xu [a], Shang Mao [b], Tao Zhou [c], Dong Wei [c] and Haolei Zhang [c]
[a] *Institute of Advanced Nuclear Energy Technology, State Power Investment Corporation Research Institute, Beijing, China,* [b] *School of Energy and Power Engineering, Zhengzhou University of Light Industry, Zhengzhou, Henan, China,* [c] *Department of Nuclear Science and Technology, Institute of Nuclear Thermal-Hydraulic Safety and Standardization, Southeast University, Nanjing, Jiangsu, China*

8.1 Future nuclear science and technology

8.1.1 Key areas of development in nuclear science and technology

8.8.1.1.1 Third generation reactors

8.1.1.1.1 Development status

As of early 2024 there are a total of 413 nuclear power units operating across 32 countries, with a combined installed capacity of 371.0 GW. In addition, 42 countries are planning to build nuclear power plants in the future (including 26 nonnuclear countries). The International Atomic Energy Agency (IAEA) predicts that by 2030, the world will see an additional 101–206 GW of new nuclear power, bringing the total installed capacity to approximately 345–554 GW [12]. In the aftermath of the Fukushima accident the international community has raised safety expectations for new nuclear power units, placing significant hope on advanced third-generation pressurized water reactors (PWRs). Among the 58 nuclear power units currently under construction in 18 countries, 49 are PWRs with an installed capacity of 52.7 GW, accounting for 88.72% of the total energy capacity; there are 3 heavy water reactors with a capacity of 1.9 GW (3.20%); 2 boiling water reactors (BWRs) with a capacity of 2.7 GW (4.55%); and 4 fast reactors with a capacity of 2.1 GW (3.54%). The third-generation nuclear power technology "Hualong One," independently developed by China, has formed a complete independent intellectual property rights and standard system. It has 41 operating and underconstruction units at home and abroad, making it the world's largest third-generation nuclear power technology in this category. The operational performance of the demonstration units has reached the international

Nuclear Science, Technology, and Civilization. DOI: https://doi.org/10.1016/B978-0-443-13941-3.00008-2
Copyright © 2026 Elsevier Inc. All rights are reserved, including those for text and data mining, AI training, and similar technologies.

top level. At the same time, it has successfully achieved technology export and entered the mass construction stage, becoming the main model of domestic nuclear power construction and the core force of nuclear power "going global."

8.1.1.1.2 Design concept and birth background

The third generation nuclear reactor does not refer to a single technology, but rather a new generation of nuclear power plant design developed on the basis of the second generation reactor (such as the PWR built in the 1970s and 1980s), drawing lessons from nuclear accidents such as Chernobyl and Three Mile Island, and incorporating the latest technological achievements. The core design concept is to prioritize safety and significantly improve the reliability and ability to withstand serious accidents of nuclear power plants through the principles of "simplification" and "nonactive" engineering, aiming to fundamentally eliminate the possibility of large-scale release of radioactive materials.

8.1.1.1.3 Economy and Standardization

Although pursuing ultimate safety, third-generation nuclear reactors are also committed to enhancing their economic competitiveness. By designing standardized and modular construction, it significantly shortens the construction cycle and reduces construction and financing costs. At the same time, its reactor power is larger, typically with a single unit capacity of over 1100 MW, higher fuel combustion efficiency, and longer refueling cycles, typically up to 18–24 months, thereby improving operational availability and economic benefits. In addition, a longer design operating life, typically 60 years, also reduces the cost of the entire lifecycle.

8.1.1.2 Small modular reactors (SMRs)

8.1.1.2.1 Ideal solution for small-scale energy adaptation

The SMR (small modular reactor) is designed for small-scale electricity or energy markets, particularly in remote regions or isolated grids where long-distance power transmission is impractical and large reactors are not feasible. SMRs can meet the flexible power generation needs in a number of fields, which include replacing decommissioned fossil fuel plants, providing heat and power for developing countries, remote areas, and off-grid locations, as well as enabling hybrid nuclear/renewable energy systems [17]. China's SMR research has formed 12 different types of reactor technologies, including the construction or commercial operation stages of multiple models such as "Linglong One," while actively promoting multiscenario applications. However, it currently faces industrialization challenges such as high costs and incomplete regulatory standards.

8.1.1.2.2 Innovative design features of SMRs

SMRs' innovative designs include inherent safety features and modular construction (integration of one or multiple reactor modules). SMRs can have multifunctional capabilities such as power generation, heating, and desalination.

Other advanced features of SMRs include high-performance fuels with increased burnup and limited expansion and fission gas release [18].

8.1.1.2.3 Innovative features driving SMR safety and efficiency

Accident-tolerant fuels are used in SMRs that can withstand high temperatures without melting and limit hydrogen generation during accidents. SMR's enhanced in-core instrumentation, offering greater accuracy and reducing conservatism in design analysis and operations. The application of digital technologies and artificial intelligence (AI) in SMRs is a key feature along with the low-pressure loop systems that use new composite materials instead of steel. An advanced concrete with high mechanical strength and impermeability is commonly used in SMRs, which makes it safer.

8.1.1.2.4 SMR innovations: Boosting economic competitiveness

These innovations significantly enhance the economic competitiveness of SMRs compared to intermittent wind and solar power, natural gas generation, or diesel generators used for specific applications, demonstration projects under construction include Argentina's CAREM-25 (integrated PWR) [19], China's HTR-PM (high-temperature gas-cooled reactor) [1,20] and Russia's KLT-40S (floating reactor) [2]. Upcoming potential projects include the US's mPower and NuScale, as well as Korea's SMART, all of which are small modular PWRs. China has introduced concepts for small PWRs, such as ACP100, CAP150, and ACPR50S, with ACP100 becoming the world's first small reactor to pass the IAEA safety review.

8.1.1.3 Fourth-generation advanced nuclear systems

8.1.1.3.1 Addressing key challenges in nuclear energy

In 2000 the United States launched the "Generation IV International Forum" (GIF) to address key challenges in nuclear energy development, including sustainability, economics, safety and reliability, waste minimization, nonproliferation, and physical protection. Six of the most promising nuclear systems were selected, including two gas-cooled reactors, two liquid-metal-cooled reactors, one supercritical water-cooled reactor [21], and one molten salt-cooled reactor [3].

8.1.1.3.2 The advanced sodium-cooled fast reactor

Among these selected reactor systems, nearly all GIF member countries believe that the advanced sodium-cooled fast reactor (SFR) [4,5], using MOX fuel, has the greatest potential to be commercially deployed within this century. The main goal of developing fast reactors is to breed nuclear fuel by inducing uranium-238 to fission or efficiently transmute it into plutonium-239 (Pu-239), which could alleviate potential shortages of natural uranium resources. SFR fuel has a higher burnup rate, allowing the fuel to remain in the reactor twice as long as in thermal reactors, and it reduces the content of minor actinides in spent fuel. SFRs can also be designed to transmute long-lived isotopes and actinides such as americium and other trans-plutonium elements.

8.1.1.4 Controlled nuclear fusion

8.1.1.4.1 Advantages and potential

Nuclear fusion is the process by which two lighter atomic nuclei combine to form a heavier atomic nucleus, releasing energy in the process. Controlled nuclear fusion refers to a regulated fusion reaction that can be sustained over time. In nature the most readily achievable fusion reaction involves the fusion of hydrogen isotopes—deuterium and tritium. This reaction has been occurring in the Sun for 5 billion years. Devices built on Earth to simulate the Sun's controlled nuclear reactions are commonly referred to as "artificial suns." The goal of controlled nuclear fusion is to achieve safe, continuous, and stable energy output, making it a potential candidate for one of the most ideal ultimate forms of energy. Compared to nuclear fission, controlled nuclear fusion offers higher energy output, abundant raw material resources, minimal radioactive waste, and greater safety. Fusion does not produce long-term, high-level nuclear radiation or the waste associated with nuclear fission, nor does it produce greenhouse gases, making it essentially pollution-free.

8.1.1.4.2 Two pathways to controlled thermonuclear fusion

Human understanding of thermonuclear fusion began with the explosion of the hydrogen bomb. Scientists have since aimed to develop a device that can effectively control the fusion process, allowing for sustained and stable energy output. There are currently two main methods for controlling nuclear fusion: inertial and magnetic confinements. Inertial confinement refers to a method where the inertia of the particles themselves is used to confine them, allowing nuclear fusion to occur. The basic idea is to use energy from a driver to form plasma from fusion fuel (deuterium and tritium) in a target capsule. In the extremely short time before the plasma particles can disperse because of inertia, they are compressed by an implosion into a high-temperature, high-density state, causing nuclear fusion to occur. Major research into this method is carried out at the US National Ignition Facility (NIF) [6,7] and China's Shenguang-III laser facility [8]. Magnetic confinement fusion uses special magnetic fields to confine the ultrahigh-temperature plasma—composed of deuterium, tritium, and free electrons—within a limited volume, enabling a controlled nuclear fusion reaction that releases significant energy [9,10].

8.1.2 Early theories and technologies in nuclear science and technology

8.1.2.1 Early nuclear reaction theories

8.1.2.1.1 Liquid drop model

In the early stages of nuclear physics research, scientists resembled the atomic nucleus to a liquid drop and used this model to explain its properties and nuclear reactions. However, with the development of quantum mechanics and nuclear force theories, it became evident that the liquid drop model could not accurately describe the complex internal structure and interactions within the nucleus. As a result, this model has gradually been phased out in modern nuclear physics.

8.1.2.1.2 Simplistic fission and fusion models

Early models of nuclear fission and fusion often overlooked the intricate structure of atomic nuclei and the various interactions that occur during these reactions. These models were sometimes too simplified to accurately predict the outcomes of nuclear reactions or the properties of reaction products. As experimental techniques improved and theoretical research deepened, more complex statistical and dynamic models were introduced to accurately describe the processes of nuclear reactions.

8.1.2.1.3 Nuclear reaction mechanism hypothesis

In the study of nuclear reactions, many hypotheses about reaction mechanisms were proposed. However, as experimental data accumulated and theoretical research progressed, some of these hypotheses were proven to be inaccurate or incomplete. For instance, early theories might have overlooked important reaction pathways or intermediate processes, leading to an incomplete understanding of reaction mechanisms. Over time, these gaps have been filled, and the related hypotheses have been either phased out or revised.

8.1.2.2 Nuclear fuel cycle technology

8.1.2.2.1 Traditional nuclear fuel disposal

Traditional "once-through" fuel approach involves directly storing or deep-burying spent nuclear fuel without reprocessing. Although this method is straightforward, it has low resource utilization and poses significant challenges for managing radioactive waste. Some countries do not consider reprocessing spent fuel from PWRs, instead opting to package or cut it into manageable pieces and store it permanently in deep geological repositories, leaving the fuel unprocessed and not fully utilized.

8.1.2.2.2 Canada's nuclear fuel cycle

Canada's nuclear power plants use heavy water reactors that are fueled by natural uranium, and after burning, the remaining uranium-235 content is only 0.3%. Therefore Canada has historically adopted a once-through approach to its nuclear fuel cycle. Other countries have also considered using the once-through strategy for their PWR nuclear fuel cycles because of the high costs associated with reprocessing and managing high-level radioactive waste.

8.1.2.2.3 Driven benefits and safety concerns

In the current global market, natural uranium prices are low, the enriched uranium market is saturated, and the price of commercial plutonium remains low and uncommercialized. For these countries the once-through approach seems more economically feasible. However, a 1000 MW PWR discharges spent fuel containing alpha-emitting actinides with radioactivity levels reaching around 100,000 curies. The long-term safety of burying such large quantities of long-lived alpha-emitting radionuclides underground is uncertain. In addition, it is not rational to dispose off large amounts of potentially usable uranium-238 and plutonium-239 by burying them without reuse.

8.1.2.2.4 Reprocessing technologies

As technology advances, more countries are turning to "reprocessing technologies," which allow the recovery of useful isotopes such as uranium and plutonium from spent fuel, enabling resource reuse and reducing the generation of high-level radioactive waste. Consequently the traditional once-through technology is likely to see a decrease in usage in new nuclear projects.

8.1.2.3 Nuclear reactor design

8.1.2.3.1 First generation

The development and construction of nuclear power plants began in the 1950s. In 1954 the Soviet Union built an experimental nuclear power plant with a capacity of 5000 kW, and in 1957, the United States built the Shippingport prototype nuclear power plant with a capacity of 90,000 kW. These achievements demonstrated the technical feasibility of nuclear power generation. Internationally, these experimental and prototype nuclear power units are classified as first-generation nuclear power plants.

8.1.2.3.2 The rise of second-generation nuclear reactors

In the late 1960s, based on experimental and prototype reactors, nuclear power units with capacities exceeding 300,000 kW, such as PWRs, BWRs, and heavy water reactors, were constructed. These units not only further confirmed the technical feasibility of nuclear power but also proved its economic viability, making nuclear power competitive with coal and hydropower. In the 1970s the energy crisis triggered by rising oil prices further accelerated the development of nuclear power. Today, most of the 400+ commercially operating nuclear reactors worldwide were built during this period. Prominent examples of second-generation reactors include the Westinghouse Model series, the System80 (1050 MWe, 2-loop PWR), and a range of BWR reactors. France's CPY and P4' reactors also belong to the Model series of standard nuclear power plants. Japan and South Korea also built several standard reactors, such as the Model 412, BWR, and System80.

8.1.2.3.3 Safety upgrades and technological progression

Second-generation nuclear plants form the backbone of the world's operational fleet. However, following accidents at Three Mile Island, Chernobyl, and Fukushima, many countries have implemented upgrades to improve the safety and efficiency of their existing reactors [13]. With advances in technology, third- and fourth-generation reactors have emerged, offering significant improvements in safety, economic efficiency, and environmental sustainability. For example, fourth-generation reactors provide enhanced safety features, longer operational lifespans, and reduced nuclear waste generation. As a result, first- and second-generation reactors are gradually being phased out for new projects, giving way to more advanced designs that align with the evolving demands of safety, efficiency, and environmental protection [14].

8.1.2.4 Nuclear material detection methods

8.1.2.4.1 Radiochemical analysis

Radiochemical analysis, also known as activation analysis, is a method that uses radioactive isotopes to analyze elements and compounds [11]. The core of this method involves transforming the target elements into radioactive isotopes through nuclear reactions and then measuring their characteristics (such as half-life or radioactivity intensity) to determine the element's quantity or presence. The process typically includes bombarding the sample with neutrons, charged particles, or gamma rays, which causes certain elements within the sample to undergo nuclear reactions and transform into radioactive isotopes. The radioactivity of these isotopes is then measured to infer the amount of the original element. This method is highly sensitive, capable of detecting extremely low concentrations of elements, and is widely used in fields such as environmental science, archaeology, and biology. However, radiochemical analysis has limitations. The process is complex, requiring specialized equipment and safety measures because of the involvement of radioactive materials. Strict adherence to safety regulations is essential, and the procedure is often time-intensive.

8.1.2.4.2 Principles and ion analysis

MS is a technique used for component analysis by measuring the mass-to-charge ratio of ions. In MS the sample is first ionized into charged particles (ions), which are then separated based on their mass-to-charge ratios as they pass through electric or magnetic fields. These ions are detected, producing a mass spectrum, from which information about the sample's molecular weight, molecular formula, and structure can be obtained. Mass spectrometry offers high sensitivity and accuracy, allowing the detection of very low concentrations of compounds while providing detailed structural information in a single analysis. It is widely applied in chemistry, biology, medicine, and environmental science. For example, in drug development, MS is used for identifying and quantifying drug components, and in environmental science, it helps detect and monitor pollutants.

8.1.2.4.3 Challenges and the rise of new detection techniques

However, MS has its own challenges, such as the need for high-purity samples and often complex pretreatment processes. In addition, operating and maintaining mass spectrometers require specialized knowledge and skills. With technological advancements, new nuclear material detection methods are emerging, such as laser-induced breakdown spectroscopy (LIBS) and neutron activation analysis [15]. These techniques offer higher sensitivity, faster detection speeds, and lower costs [16]. As a result, traditional radioactive detection methods are likely to be gradually replaced by newer techniques in both new projects and system upgrades.

8.2 A better "nonnuclear" world

8.2.1 How nuclear-free countries view a "nuclear-free" world

8.2.1.1 Perspectives of nuclear-free countries on nuclear issues

8.2.1.1.1 Vigilance and concern about nuclear weapons

Nuclear-free countries generally hold a cautious attitude toward nuclear weapons, viewing them as highly destructive and a significant threat to humanity. These countries are deeply concerned about the potential for nuclear war and the broader impact such a conflict would have on global peace and security. For nations like Japan, which suffered the devastating effects of nuclear attacks during World War II, the commitment to nuclear disarmament is not only an international expectation but also a domestic demand. Since the 1970s, Japan has adhered to its "three non-nuclear principles," which state that Japan shall not possess, produce, or permit the introduction of nuclear weapons into its territory. This policy reflects Japan's profound understanding of the dangers of nuclear weapons and its strong commitment to peaceful development. For Japan, being a nuclear-free state is both a pledge to the international community and a cornerstone of domestic political stability and public safety. Japan's rightward shift has intensified, and some politicians have proposed building an Asian version of NATO, attempting to introduce US nuclear weapons. Their nuclear ambitions have sparked strong opposition from countries such as China, the United States, and Russia, all of which have stated that they will never allow Japan to possess nuclear weapons. South Korea also hopes to acquire nuclear submarine technology by joining the United States–United Kingdom–Australia "AUKUS" alliance. Its nuclear activities have raised concerns in the United States, and it has been listed as a "sensitive country" by the United States. Australia, South Korea, and Japan have also begun developing nuclear submarines. In October 2025 Trump announced on social media that he had instructed the Department of Defense to initiate reciprocal nuclear weapons testing.

8.2.1.1.2 Support for nuclear nonproliferation

Nuclear-free countries typically advocate for the international nuclear nonproliferation regime, viewing it as a crucial mechanism for preventing the spread of nuclear weapons and maintaining international peace and security. These nations call for stronger global cooperation to combat nuclear terrorism, prevent the illicit transfer of nuclear materials, and ensure that nuclear technology is not misused for military purposes.

8.2.1.1.3 Expectations for the peaceful use of nuclear energy

Although nuclear-free countries do not possess nuclear weapons, many support the peaceful use of nuclear energy. They view nuclear power as a clean and efficient energy source that can significantly contribute to economic growth and societal progress. For these countries, the peaceful application of nuclear technology, especially in the energy sector, is seen as a valuable tool for achieving sustainable development while addressing the challenges of climate change. Avoiding nuclear weapons is the core consensus of the international community to maintain global strategic stability and prevent nuclear proliferation risks, which needs to be promoted through legal constraints, mechanism supervision, and diplomatic cooperation in multiple dimensions.

8.2.1.2 Impacts of a "nuclear-free" world on nuclear-free countries

8.2.1.2.1 Changes in international status

If the world achieves nuclear disarmament, nuclear-free countries may see an elevation in their international standing. This is because these countries play an important role in the global nuclear nonproliferation and disarmament processes, and their efforts and contributions will be recognized and appreciated by the international community. In addition, nuclear-free countries may gain greater voice and influence in international affairs.

8.2.1.2.2 Impact on development

Achieving a nuclear-free world will help nuclear-free countries establish a more stable and secure environment for development. Without the threat of nuclear weapons, these countries can focus more on economic growth, social progress, and improving the well-being of their citizens. Furthermore a nuclear-free world will promote cooperation and exchanges between countries, providing nuclear-free nations with more opportunities for development and collaboration. However, it is important to note that achieving nuclear disarmament is not an overnight process; it requires collective effort and long-term cooperation from the international community. Throughout this process, nuclear-free countries should continue to uphold the principles of peace, cooperation, and mutual benefit, actively participating in global nonproliferation and disarmament initiatives, and making greater contributions to the advancement of global peace and security.

8.2.2 Nuclear-armed countries and "nuclear-free" world

8.2.2.1 Views of nuclear-armed countries on nuclear weapons

8.2.2.1.1 Strategic deterrence role of nuclear weapons

Nuclear-armed countries generally view nuclear weapons as an essential component of national strategic deterrence. As an extreme military means, nuclear weapons are seen as a final line of defense that can ensure national security, prevent external aggression, and protect national interests.

8.2.2.1.2 Military advantages of nuclear weapons

Possessing nuclear weapons is considered a means of gaining an advantageous position in international military competition. Nuclear-armed countries may enhance their military strength and strategic influence through nuclear weapons, thereby gaining greater voice and leadership in international affairs.

8.2.2.1.3 Peaceful use of nuclear energy

Beyond military applications, nuclear-armed countries also emphasize the peaceful use of nuclear energy. They may believe that nuclear energy is an efficient and clean energy source, playing a significant role in promoting economic development, improving energy structures, and addressing climate change.

8.2.2.1.4 Positions on nonproliferation and disarmament

Nuclear-armed countries hold differing positions on nuclear nonproliferation and disarmament issues. Some may advocate for promoting global nuclear disarmament and nonproliferation processes through international cooperation, whereas others may focus more on the security of their own nuclear arsenals and strategic interests.

8.2.2.2 Impact of "nuclear-free" world on nuclear-armed countries

8.2.2.2.1 Changes in international status

If global nuclear disarmament is achieved, the international status of nuclear-armed countries may be affected to some extent. On one hand, these countries would lose an important strategic deterrent, potentially putting them at a relative disadvantage in international military competition; on the other hand, nuclear disarmament could foster greater international cooperation and exchange, providing nuclear-armed countries with more development opportunities and cooperation space. Setting an example in the context of "nonnuclear" means that nuclear countries take practical actions to promote nuclear disarmament, whereas nonnuclear countries adhere to their commitment not to possess nuclear weapons, providing a reference paradigm for global nuclear governance.

8.2.2.2.2 Military strategy adjustments

Nuclear disarmament would prompt nuclear-armed countries to reassess their military strategies and defense systems. These countries might need to adjust their military deployments, strengthen conventional military capabilities, and enhance intelligence, surveillance, and reconnaissance capabilities to address potential security threats postnuclear disarmament.

8.2.2.2.3 Domestic political and economic impact

Nuclear disarmament would have profound implications for the domestic politics and economies of nuclear-armed countries. On one hand, these countries would need to tackle the economic and technical challenges posed by nuclear weapons reduction and elimination; on the other hand, nuclear disarmament would also drive these countries to strengthen international cooperation and exchange, promoting economic development and social progress.

8.3 A better nuclear world

8.3.1 A better concept of nuclear safety

8.3.1.1 Nuclear safety concepts of other major nuclear-armed countries

8.3.1.1.1 United States

The United States has long regarded nuclear safety as a vital component of national security and has implemented a series of measures to strengthen nuclear safety. For instance the United States enhances the management and regulation of nuclear materials through legislation and promotes international cooperation to combat nuclear terrorism.

8.3.1.1.2 Russia

Russia also places a high priority on nuclear safety, viewing it as a cornerstone of national security. The country has taken several measures to enhance the safety and regulation of nuclear facilities and actively participates in international nuclear safety cooperation.

8.3.1.1.3 France

As one of Europe's leading nuclear energy countries, France maintains a strong nuclear safety perspective. France ensures the safe operation of nuclear facilities by establishing a robust regulatory framework and technical standards while actively engaging in international nuclear safety cooperation and dialogue.

8.3.1.1.4 United Kingdom

The UK government emphasizes the importance of nuclear safety and has implemented various measures to enhance the security and regulation of nuclear facilities. In addition, the United Kingdom is actively involved in international nuclear safety cooperation, contributing to the improvement of the global nuclear safety governance system.

8.3.1.1.5 Japan

Japan is not a nuclear-weapon state and adheres to the "three non-nuclear principles," which entail not possessing, not manufacturing, and not introducing nuclear weapons. Japan's nuclear safety philosophy primarily revolves around the peaceful use of nuclear energy, emphasizing the rational and appropriate utilization of nuclear energy based on a clear responsibility system and effective risk management and control. Its goal is to achieve the trust of its citizens under the premise of safety, with the aim of peaceful use, and to contribute to the environment, people's lives, and economic development through the development of nuclear energy technology.

8.3.1.1.6 China

China and other major nuclear-armed nations share a strong commitment to nuclear safety and have implemented various measures to enhance it. These measures include strengthening domestic nuclear safety regulatory frameworks and technical standards, promoting international cooperation to combat nuclear terrorism, and enhancing the security and regulation of nuclear facilities. Concurrently, countries actively participate in international nuclear safety cooperation and dialogue, collectively advancing the improvement and development of the global nuclear safety governance system.

8.3.1.2 Chinese concept of nuclear safety

8.3.1.2.1 Balancing development and security

China believes that nuclear energy is a crucial avenue for ensuring energy security and addressing climate change. The peaceful use of nuclear energy ignites hope for human development. However, without effective nuclear safety measures, this promising outlook could be overshadowed. Therefore China upholds the principle of seeking security for development and promoting development through security, ensuring the growth of nuclear energy is predicated on safety.

8.3.1.2.2 Balancing rights and obligations

China calls on more nations to actively consider ratifying international legal documents such as the Convention on the Physical Protection of Nuclear Material and its amendments, as well as the International Convention for the Suppression of Acts of Nuclear Terrorism. At the same time, China respects the rights of countries to adopt nuclear safety policies and measures that best fit their national circumstances, jointly advancing the international nuclear safety process.

8.3.1.2.3 Balancing autonomy and cooperation

China asserts that nuclear safety is primarily a national issue, with the primary responsibility resting on individual governments. However, it is also a global concern that requires collaborative efforts. China is willing to enhance communication, mutual learning, and sharing with all nations to jointly pursue universal nuclear safety.

8.3.1.2.4 Balancing immediate and fundamental solutions

China emphasizes that improving nuclear safety policies and measures, developing modern and low-risk nuclear technologies, strengthening nonproliferation export controls, and deepening international cooperation against nuclear terrorism are direct and effective ways to eliminate nuclear safety hazards and proliferation risks. In addition, fostering a peaceful and stable international environment is fundamental for the lasting safety and development of nuclear energy. Furthermore, China actively promotes international cooperation in nuclear safety, sharing technology and experiences, contributing resources and platforms, and is committed to building a fair, cooperative, and win–win international nuclear safety system.

8.3.1.3 Evaluation

In the future, as the global nuclear energy industry continues to evolve and international circumstances change, nations need to strengthen cooperation and communication to address nuclear safety challenges and promote the ongoing enhancement and development of the global nuclear safety governance system.

8.3.2 A better nuclear safety regulatory framework

8.3.2.1 Nuclear safety regulatory frameworks of other major nuclear-armed countries

8.3.2.1.1 United States

The core of the legal framework is the Atomic Energy Act, enacted in 1954. This law establishes the legal foundation for nuclear energy development in the United States and outlines a comprehensive framework for its management. The nuclear facility safety regulations detail safety requirements and management measures for nuclear facilities. The Nuclear Waste Policy Act addresses issues related to the handling, storage, and disposal of nuclear waste. The regulatory framework is primarily overseen by the Nuclear Regulatory Commission (NRC), which is responsible for the safety regulation of civilian nuclear facilities and activities, whereas Congress designates some nonmilitary nuclear facilities and activities under the jurisdiction of the Department of Energy to the NRC. The regulatory types include policies, orders, notifications, manuals, guidelines, and technical standards.

8.3.2.1.2 Russia

The legal core is the law about the utilization of atomic energy, which outlines the fundamental principles, regulatory bodies, and responsibilities for the use of atomic energy. The law about state policy in the field of radioactive waste management sets forth policies, principles, and measures for managing radioactive waste. The regulatory model follows a centralized management approach, with the Ministry of Atomic Energy overseeing nuclear military research and production, as well as nuclear energy technology development. The regulatory framework consists of a legal structure established within the Russian Federation's legal system, including the constitution, presidential decrees, and State Duma laws. Specialized agencies, such as the Federal Nuclear Regulatory Agency and the Facilities Safety Committee, are responsible for specific nuclear safety oversight.

8.3.2.1.3 United Kingdom

Major regulations include the "Ionizing Radiations Regulations" and the "Radioactive Substances Act." The regulatory framework continuously adapts and modifies relevant regulations in line with advancements in nuclear technology, delineating the roles of the Ministry of Defense in overseeing military nuclear operations and the Department for Business, Energy & Industrial Strategy in managing civilian nuclear operations. The Defense Nuclear Safety Regulator (DNSR) is responsible for overseeing the safety of military nuclear facilities and activities, whereas the Office for Nuclear Regulation (ONR) manages the safety of civilian nuclear facilities and activities.

8.3.2.1.4 France

The legal core is the "Transparency and Security in the Nuclear Field Act," which establishes basic principles and requirements for nuclear energy use, facility safety, and waste management. Much of the legislation is derived from international recommendations and regulations, such as those from the International Commission on Radiological Protection and European Union directives. The Anti-Pollution Law outlines fundamental principles and measures for pollution prevention, including controls on nuclear contamination. The comprehensive protection of individuals against ionizing radiation details protective measures and regulatory requirements related to ionizing radiation. The regulatory framework is overseen by the Nuclear Safety Authority (ASN), which is responsible for nuclear safety and radiation protection. The regulations are categorized into several areas, including radiation safety, operational safety, protection, physical protection, and transportation.

8.3.2.2 China's nuclear safety regulatory framework

8.3.2.2.1 Development history

In 1982 the Ministry of Nuclear Industry began organizing research and adapting nuclear power plant safety regulations from the IAEA. The State Council successively approved and issued several regulations, including the Regulations on the Safety Supervision and Administration of Civil Nuclear Facilities of the People's Republic of China (1986), the Regulations on the Management of Nuclear Materials of the People's Republic of China (1987), and the Regulations on Emergency Management of Nuclear Accidents in Nuclear Power Plants (1993). Based on these regulations, China established a series of nuclear safety rules and guidelines. The Guidelines for Nuclear Safety Reporting of Research Reactor Operating Units implemented on April 1, 2025, have refined the safety reporting requirements for research reactors. In addition, the Ministry of Ecology and Environment will strengthen nuclear and radiation safety supervision in 2023, including issuing nuclear power construction permits and promoting the retirement of old nuclear facilities. The Atomic Energy Law of the People's Republic of China, issued in September 2025, stipulates that when engaging in atomic energy research, development, and utilization activities, we should adhere to the leadership of the Communist Party of China (CPC), implement the overall national security concept, and adhere to a rational, coordinated, and progressive nuclear security concept. Although emphasizing the development of nuclear fusion, especially the deuterium tritium route, China has begun to continuously improve the regulatory system for nuclear fusion experiments.

8.3.2.2.2 Framework composition

The latest Energy Law of the People's Republic of China, implemented on September 12, 2025, specifies the following key contents in terms of nuclear safety: Actively develop nuclear power in a safe and orderly manner, strengthen the nuclear safety responsibility system, align technical standards with international standards, promote safe use of nuclear energy and public participation, establish regulatory mechanisms, and promote technological innovation. The Energy Law, guided by the principles of "safety, reliability, green, and low-carbon," has established a legal framework for the development of nuclear power through full process supervision, an improved responsibility system, and upgraded technical standards and public participation mechanisms, providing important guarantees for China to achieve its carbon peak and carbon neutrality goals. The Atomic Energy Law, the Nuclear Safety Law, and the Law on the

Prevention and Control of Radioactive Pollution jointly constitute the legal cornerstone of China's nuclear energy and nuclear safety governance. Each has its own focus and is interconnected. The three work together to form a three-in-one legal protection system of "development safety environmental protection," jointly supporting the safe and sustainable development of China's nuclear energy and nuclear technology.

8.3.2.3 Better progress in domestic and international nuclear safety supervision

Significant progress has been made in regulatory standards, technical means, regulatory models, and international cooperation in nuclear safety supervision both domestically and internationally, forming a more rigorous, efficient, and collaborative global nuclear safety governance pattern. On the foreign front the US NRC is promoting digital transformation and the application of AI, optimizing the regulatory "sunset" mechanism to adapt to the development of advanced reactors; France strengthens risk guidance and supervision throughout the entire chain, whereas the EU improves cross-border emergency linkage; Russia has upgraded its regulatory system through legislative revisions, and countries have generally improved the efficiency and safety requirements for the review of new technologies such as small reactors, ensuring national nuclear safety. A "highest and strictest" regulatory standard system centered on the Nuclear Safety Law has been established domestically, forming a trinity regulatory model and a full process and full chain supervision mechanism for the Atomic Energy Law, Nuclear Safety Law, and Radioactive Pollution Prevention and Control Law. With the help of AI and big data an intelligent regulatory platform is created, and the radiation monitoring network is fully covered. At the same time, it deeply participates in international cooperation such as the IAEA, shares regulatory practices, and promotes the construction of standard discourse power, which has been highly recognized by international evaluations for national nuclear safety. At the international level the IAEA Nuclear Security Action Plan promotes global peer review and emergency response coordination, strengthens bilateral and multilateral cooperation among multiple countries, jointly addresses nuclear security challenges, and builds a strong global nuclear security defense line.

8.3.3 A better nuclear safety environment

8.3.3.1 The importance of human factors in nuclear safety

8.3.3.1.1 Personnel selection and training

Rigorous selection processes are in place to ensure that only individuals with the necessary professional skills and sound psychological qualities are allowed to work in nuclear power plants. Regular training programs are conducted to enhance employees' professional skills and nuclear safety awareness. Emphasis is placed on teamwork and communication to ensure personnel can make prompt and accurate decisions in emergency situations.

8.3.3.1.2 Optimizing the working environment

Key areas, such as control rooms, are designed with consideration for the physiological and psychological characteristics of personnel, creating a comfortable work environment. Human-machine interface designs are optimized to ensure that personnel can access information accurately and respond swiftly.

8.3.3.1.3 Accident simulation and emergency drills

Accident simulations and emergency drills are conducted to improve personnel's response capabilities in the event of a nuclear incident. The causes of errors during accidents are analyzed to learn from experiences and continuously improve nuclear safety management measures.

8.3.3.2 Building a nuclear safety culture

8.3.3.2.1 Leadership exemplification

Decision-makers and management demonstrate their commitment to nuclear safety through tangible actions, setting a positive example for employees. They actively engage in promoting nuclear safety culture and fostering a participatory atmosphere among all staff.

8.3.3.2.2 System development

Establishing and perfecting nuclear safety management systems that clarify responsibilities and obligations. Implementing strict reward and punishment mechanisms to encourage compliance with nuclear safety regulations and impose severe penalties for violations.

8.3.3.2.3 Training and public awareness

Conducting regular nuclear safety training enhances employees' awareness and skill levels regarding nuclear safety. Utilizing various channels to disseminate nuclear safety knowledge and improve public understanding and awareness.

8.3.3.2.4 Experience feedback and continuous improvement

Creating a feedback mechanism to summarize and analyze lessons learned from nuclear safety incidents promptly. Continuously improving nuclear safety management measures based on feedback results to enhance safety levels.

8.3.3.2.5 International communication and cooperation

Actively participating in international nuclear safety exchanges and collaborations to learn from advanced experiences and technologies in other countries.

Collaborating to address nuclear safety challenges and promoting the improvement and development of global nuclear safety governance systems.

8.3.3.2.6 Policy guidance

Government agencies have issued a series of policy documents regarding nuclear safety culture, providing strong support for its development. The core principle of "nuclear safety is paramount" is vigorously advocated, embedding nuclear safety culture into the fabric of society.

8.3.3.2.7 Public participation

Organizing various nuclear science popularization activities to enhance society's understanding and recognition of nuclear safety. Establishing a mechanism for broad public participation to solicit feedback on significant nuclear safety matters that affect public interests.

8.3.3.3 Future development direction

8.3.3.3.1 Humanities and high technology development

The humanistic and high-tech development of nuclear energy is a reflection of the deep integration of technological progress and human social needs. From a technical perspective, nuclear energy relies on cutting-edge technologies such as nuclear physics research, nuclear reactor design, and nuclear fuel cycle, constantly breaking through the technological boundaries of energy conversion efficiency and safety control. For example, the innovation of the fourth-generation nuclear reactor in inherent safety and miniaturization promotes the extension of nuclear energy from single power generation to multiple scenarios such as industrial heating and seawater desalination. From a humanistic perspective the development of nuclear energy has always revolved around the core of "putting people first." It is necessary to support climate change, ensure energy accessibility, and provide support for sustainable development of human society through stable low-carbon energy supply. It is also necessary to balance the public's right to know and participate, eliminate cognitive misconceptions through science popularization, balance technological progress and social acceptance, and achieve the unity of technological value and humanistic care.

8.3.3.3.2 Development of intelligent and new quality productivity

The development of nuclear energy intelligence and new quality productivity is a new development paradigm born from the deep integration of digital technology and the nuclear energy industry. At the level of intelligence, through technologies such as AI, big data, and the Internet of Things, the nuclear energy field has achieved real-time monitoring of nuclear reactor status, fault warning, and intelligent regulation. For example, AI algorithms optimize core fuel management, and digital twin technology simulates the full life cycle operation of nuclear facilities, greatly improving operation and maintenance efficiency and safety levels; at the level of cultivating new quality productivity, the intelligentization of nuclear energy promotes the reconstruction and value upgrading of the industrial chain. On the one hand, it gives birth to new industrial forms such as nuclear safety monitoring equipment and intelligent nuclear fuel components. On the other hand, it expands application scenarios through the "nuclear energy+" model, such as providing customized energy solutions for remote industrial and offshore platforms through intelligent SMRs, driving the upgrading of related industries such as high-end equipment manufacturing and software services, and forming a new productivity growth pole with technological innovation as the core and efficiency improvement as the guide.

8.3.3.3.3 Safe and civilization development

The safe and civilized development of nuclear energy is an energy answer sheet intertwined with a beautiful vision and intelligent technology. From a beautiful perspective, it is a "green engine" that safeguards the ecology—replacing high-polluting energy sources for power generation, gradually dispersing smog over cities, and keeping the water quality of rivers, lakes, and seas clean; going deeper into the details of people's livelihoods, using stable energy to support winter heating and medical equipment operation, making homes warm in the cold winter, making cancer radiotherapy and other diagnostic and treatment technologies easier to popularize, quietly painting a picture of livable life. This beauty is built on the foundation of intelligent technology. Nuclear reactors are equipped with intelligent sensing systems that can capture real-time temperature, pressure, and other data to accurately warn of risks; AI algorithms can simulate extreme working conditions and optimize safety plans in advance; the digital operation and maintenance platform realizes full process visualization management, making every step of the operation controllable and traceable. Human beings use intelligent technology to equip nuclear safety with a "smart brain," which not only fully unleashes the clean value of nuclear energy but also firmly holds the safety red line, making it a reliable driving force for promoting civilization progress. In the rapid advancement of AI today, nuclear energy stands as the sole viable and reliable power source for AI. The development of nuclear technology paints the most brilliant and beautiful foundation for human intellectual civilization.

References

[1] Y. Wu, et al., Corrigendum to NECP-Panda: a neutronics and thermal-hydraulic analysis code for pebble-bed high temperature gas-cooled reactor, Ann. Nucl. Energy 210 (2025) 110866.

[2] S.V. Beliavskii, V.N. Nesterov, R.A. Laas, et al. Effect of fuel nuclide composition on the fuel lifetime of reactor KLT-40S-ScienceDirect. Nucl. Eng. Des., 360. doi:10.1016/j.nucengdes.2020.110524.

[3] C. Wang, Q. Duan, M. Liu, et al., Code development and analysis of heat pipe cooled passive residual heat removal system of Molten salt reactor, Ann. Nucl. Energy 144 (2020) 107527, doi:10.1016/j.anucene.2020.107527.

[4] Z. Liu, S. Cheng, W. Li, et al., Review of researches on flow blockage accident in liquid-metal-cooled reactor, Prog. Nucl. Energy 176 (2024), doi:10.1016/j.pnucene.2024.105372.

[5] K.M. Sourov, M.H. Sahadath, H.R.K. Rohan, Enhancing fuel breed-burn performance in a sodium-cooled fast reactor using a novel reactivity control method, Prog. Nucl. Energy 177 (2024), doi:10.1016/j.pnucene.2024.105436.

[6] O.A. Hurricane, How ignition and target gain >1 were achieved in inertial fusion, High Energy Density Phys. 53 (2024), doi:10.1016/j.hedp.2024.101157.

[7] K.L. Baker, P.A. Amendt, D.A. Mariscal, et al., Frustraum 1100 experimental campaign on the National Ignition Facility, High Energy Density Phys. 53 (2024), doi:10.1016/j.hedp.2024.101158.

[8] J. Wen, S. Peng, X. He, H. Li, P. Lu, Engineering Application of Fusion Reactors in China: Problems and Strategies, Strategic Study of CAE 27 (6) (2025) 192–205.

[9] L. Alessandro, M. Giuseppe, A. Franco, et al., Progress of the plasma centerpost for the PROTO-SPHERA spherical tokamak, Energies 9 (7) (2016) 508, doi:10.3390/en9070508.

[10] V.M. Sharapov, Discharge chamber plasma-chemical conditioning in magnetic confinement fusion devices (Review), Phys. At. Nucl. 84 (7) (2021) 1266–1271, doi:10.1134/S1063778821070139.

[11] Y. jingjing, et al., General methods, application development and related problems of radiochemical purity analysis, J. Isot. 36 (04) (2023) 380–388.

[12] R. Timakova, R. Iliukhin, I. Iliukhina, Modern trends in the development of nuclear power: from environmental friendliness to safe utilitarianism, IOP Conf. Ser.: Earth Environ. Sci. 979 (2022), doi:10.1088/1755-1315/979/1/012172.

[13] N. Kinoshita, K. Sueki, K. Sasa, J.I Kitagawa, et al., Assessment of individual radionuclide distributions from the Fukushima nuclear accident covering central-east Japan, Proc. Nat. Acad. Sci. U.S.A. 108 (49) (2011) 19526–19529, doi:10.1073/pnas.1111724108.

[14] H. Yu, B. Tang, Y. Fu, et al., Quantifying the reproducibility and longitudinal repeatability of radiomics features in magnetic resonance image-guide accelerator imaging: a phantom study, Eur. J. Radiol. 181 (2024), doi:10.1016/j.ejrad.2024.111735.

[15] I.V. Malyshev, N.I. Chkhalo, S.N. Yakunin, Project of X-ray optical scheme of a lithograph with a transmissive dynamic mask and a synchrotron radiation source, Tech. Phys. 69 (7) (2024), doi:10.1134/S1063784224070272.

[16] A.J Bevan, F.E Taifakou, C.Z Amjad, et al., Organic electronic-based neutron detectors, Nucl. Instrum. Methods Phys. Res. A 1069 (2024), doi:10.1016/j.nima.2024.169882.

[17] E.M.A. Hussein, Emerging small modular nuclear power reactors: a critical review, Phys. Open 5 (2020) 100038, doi:10.1016/j.physo.2020.100038.

[18] E.J.T. Berg, A. Buijs, Evaluation framework for molten salt reactors and other new nuclear power reactor systems, Nucl. Eng. Des. 429 (2024), doi:10.1016/j.nucengdes.2024.113588.

[19] S.Z. Ganjaroodi, M. Fani, E. Zarifi, et al., Thermal-hydraulic modeling of CAREM-25 advanced small modular reactor using the porous media approach and COBRA-EN modified code, Nucl. Eng. Technol. 56 (5) (2024) 1574–1583, doi:10.1016/j.net.2023.12.011.

[20] Z. Zhang, Q. Zou, N. Gui, et al., Pebble flow in the HTR-PM reactor core by GPU-DEM simulation: effect of friction, Nucl. Eng. Technol. 56 (9) (2024) 3835–3850, doi:10.1016/j.net.2024.04.032.

[21] H. Han, C. Zhang, Construction of the dynamic model and control system for the Canadian supercritical water–cooled reactor power plant, Nucl. Technol. 210 (5) (2024) 836–849, doi:10.1080/00295450.2023.2249710.

Index

Page numbers followed by "*f*" and "*t*" indicate, figures and tables respectively.

H

I

J

K

L

M

O

P

Printed and bound by CPI Group (UK) Ltd, Croydon, CR0 4YY
29/08/2026
02206142-0001